Analytical
Biochemistry
of Insects

Analytical Biochemistry of Insects

EDITED BY

Ralph B. Turner

DEPARTMENT OF BOTANY AND ENTOMOLOGY
NEW MEXICO STATE UNIVERSITY, LAS CRUCES, N.M.

ELSEVIER SCIENTIFIC PUBLISHING COMPANY
AMSTERDAM − OXFORD − NEW YORK 1977

ELSEVIER SCIENTIFIC PUBLISHING COMPANY
335 Jan van Galenstraat
P.O. Box 211, Amsterdam, The Netherlands

Distributors for the United States and Canada:

ELSEVIER/NORTH-HOLLAND INC.
52, Vanderbilt Avenue
New York, N.Y. 10017

Library of Congress Cataloging in Publication Data

Main entry under title:

Analytical biochemistry of insects.

    Includes index.
    1. Insects--Phyiology. 2. Biological chemistry--
Technique. I. Turner, Ralph B. [DNLM: 1. Insects--
Analysis. 2. Invertebrate hormones--Analysis.
3. RNA--Analysis. QX500 A532]
QL495.A5   595.7'01'9285   76-54362
ISBN 0-444-41539-4

Printed in The Netherlands

CONTENTS

LIST OF CONTRIBUTORS

M. T. Armold, Department of Chemistry, Montana State University, Bozeman, Montana 59715 U.S.A.

P. S. Chen, Institute of Zoology, University of Zurich, Zurich, Switzerland

F. L. D Lucca, Departmento de Bioquimica, Faculdade de Medicina de Ribeirao, Preto da Universidade de Sao Paulo, Ribeirao/Preto, Sao Paulo, Brazil

L. L. Jackson, Department of Chemistry, Montana State University, Bozeman, Montana 59715 U.S.A.

M. Koreeda, Department of Chemistry, The Johns Hopkins University, Baltimore, Maryland 21218 U.S.A.

C. A. Lang, Department of Biochemistry and the Biomedical Aging Research Program, University of Louisville School of Medicine, Louisville, Kentucky 40201 U.S.A.

W. Mastropaolo, Department of Biochemistry and the Biomedical Aging Research Program, University of Louisville School of Medicine, Louisville, Kentucky 40201 U.S.A.

B. J. Mills, Department of Biochemistry and the Biomedical Aging Research Program, University of Louisville School of Medicine, Louisville, Kentucky 40201 U.S.A.

N. Mitlin, Boll Weevil Research Laboratory, United States Department of Agriculture, Agricultural Research Service, Mississippi State, Mississippi 39762 U.S.A.

D. A. Schooley, Zoecon Corporation, 975 California Avenue, Palo Alto, California 94304 U.S.A.

T. Smyth, Jr., Department of Entomology, Department of Biochemistry and Biophysics, The Pennsylvania State University; 2 Patterson Building, University Park, PA 16802 U.S.A.

B. A. Teicher, Department of Chemistry, The Johns Hopkins University, Baltimore, Maryland 21218 U.S.A.

B. N. White, Department of Biology, Queen's University, Kingston, Canada K7L 3N6

PREFACE

Although there is an abundance of books describing analyses of biochemicals, the peculiarities of the composition of insect tissue present frustrating obstacles to the valid application of many methods routinely used with microorganisms or vertebrate tissues. This book presents the successful adaptation of such methods to insect tissues. The authors have drawn extensively on their own personal experience as well as literature references in order to fully describe applications, and to present methods in sufficient detail to permit the novice and experienced researcher alike to accomplish the rigor demanded for modern biochemical interpretations.

Analyses of carbohydrates are not included because no unique problems have been encountered in the analysis of insect carbohydrates by routine methods. However, it should be noted that the "essential conditions of vigorous analytical methodology" (C. A. Lang, et al. p. 43) must be met for analysis of carbohydrates as for other biochemicals.

I am indebted to the authors of this volume for their cooperation and their individual care in preparation with regard to quality and accuracy of content.

Ralph B. Turner
Professor of Insect Biochemistry
Department of Botany and Entomology
New Mexico State University
Las Cruces, New Mexico   88003

CHAPTER 1

THE ANALYSIS OF NUCLEOSIDES, NUCLEOTIDES, AND ASSOCIATED COMPOUNDS

NORMAN MITLIN

Boll Weevil Research Laboratory, U. S. Department of
Agriculture, Agric. Res. Serv., Mississippi State, MS 39762
(U.S.A.)

CONTENTS

INTRODUCTION

Although the literature is replete with methods for analyzing nucleosides, nucleotides, and related compounds, relatively few such methods seem to have been devised or adapted for
use with insects. It is the rare insect biochemist or physio-
logist who has not had the unrewarding experience of strug-
gling unsuccessfully with a published analytical method
devised for some other organism.

In this chapter, therefore, we report only those analytical
methods that have been demonstrated to be effective in work

Abbreviations used: Ado = adenosine; ADP = adenosine diphos-
phate; AMP = adenosine monophosphate; ATP = adenosine triphosphate; Cyt = cytosine; de Ado = deoxyadenosine; de Guo =
deoxyguanosine; deo Urd = deoxyuridine; Guo = guanosine; GMP =
guanosine; GMP = guanosine monophosphate; G 6PDH = glucose-6-
phosphate dehydrogenase; HK = hexokinase; HMP = hexomonophos-
phate; Methyl de Cyt = methydeoxycytosine; 15 pseudo-Urd =
pseudouridine; Thy = thymidine; TPN = triphosphopyridine nucleotide; TPNH = triphosphopyridine nucleotide, reduced; Urd =
uridine.

with insects, though many methods which are effective with
other organisms could be of value to those working with insects.
Also, because laboratories differ in the kinds of equipment
available to them, we have selected methods, where possible,
that can be used by investigators in different situations.
Obviously this is not a full compendium of procedures, but
those methods selected have been proved successful with at
least one species of insect.

SAMPLE PREPARATION - GENERAL PROCEDURES

The investigator who sets out to prepare a sample of tissue
for study is confronted at once by the necessity of preparing
the material in such a way as to minimize degradation or other
biological changes due to excission or enzymatic action. The
matter is particularly sensitive when cyclic nucleotides are
to be analyzed. Most investigators have dealt with this problem by freezing the tissues immediately with some type of gas
with a low boiling temperature such as liquid nitrogen or
dichlorodifluoromethane (Freon 12) [1]. Both are readily
available. To obtain the latter as a liquid, one inverts the
can of refrigerant (CCl_2F_2) and dispenses the liquid into a
small beaker that is then suspended in a nitrogen-filled Dewar
flask. The tissue samples are immersed in the freezing CCl_2F_2
and then should be stored at below $-35^{\circ}C$ [2]. Also, the cited
authors recommend that all dissections and weighing be done in
a room maintained at $-20^{\circ}C$.

For extraction perchloric acid is the choice of many investigators [3,4,5] and others . The concentrations used are 0.4
to 0.6 N. The tissues are ground in the cold in a glass

homogenizer and centrifuged 3 to 4 times in the cold. Then the supernatants are combined and neutralized with a strong solution of KOH. The resulting $KClO_4$ is removed by centrifugation. If separation of nucleotides is to be done by anion exchange methods, the solution should be treated with activated charcoal under acid conditions. Excess electrolytes are thus removed, and the nucleotides may be eluted with either 1% ammonia in 60% ethanol or 10% pyridine in 50% ethanol [6]. The solution containing the nucleotides is now ready for separation.

Tissues prepared for radioimmunoassay, as with Drosophila melanogaster [7], are homogenized in 1N $HClO_4$. Then they are centrifuged at 15,000 rev/min for 5 min, and the precipitated proteins and other insoluble matter are discarded. The supernatant is neutralized with K_2CO_3(6.6M), and the **precipitate** is discarded after centrifugation. The nucleotides in the supernatant are then succinylated by adding 5 mg succinic anhydride and 10 μl of triethylamine to 100 ml of supernatant. The mixture is shaken vigorously for 10 min.

For isolation of cyclic nucleotides, Fallon and Wyatt, in their work with Antheraea pernyi [8], prepared their samples differently. They homogenized the tissues at $0^{o}C$ in ethanol-HCl (60 parts absolute EtOH-1 part HCl) and after centrifugation at 12,000 g for 10 min evaporated the supernatant at room temperature under vacuum. Lipids were removed by suspending the tissue residue in 1.0 ml of 50 mM sodium acetate (pH 4.0) and extracting with water-saturated ether. The ether was then removed and the aqueous sample was ready for use.

ANALYSIS OF NUCLEOSIDES

Studies of insect nucleosides appear to be limited.   In
the few reported, the researchers used either paper or thin-
layer methods to isolate and identify the nucleosides.   For
example, in a series of studies involving the biosynthesis of
DNA in Hyalophora cecropia [9,10,11], various nucleosides were
isolated by using paper chromatography.   The same type of
procedure was used by Forest et al. [12] in work with the
milkweed bug, Oncopeltus fasciatus.

Extraction procedures

In addition to the procedures detailed, several additional
procedures have been used: (1) Eggs may be squashed directly
onto chromatographic paper [12], or methanol-pyridine-water
(5:1:4 v/v) may be used to extract tissues, in which instance
the extract should be concentrated before use.

Nucleosides may be isolated directly from an enzymatic
reaction mixture consisting of 0.2 ml of Cecropia homogenate
(in water); tris buffer (pH 7.5) 20 μmoles; ATP 1.5 μmoles;
creatine phosphate 8.0 μmoles; creatine phosphokinase 10.0 μg;
$MgCl_2 \cdot H_2O$ 8 μmoles.   70% methanol is used to stop the reaction
[9].

Separation methods

Paper chromatography

(1)   Whatman #3 chromatography paper.   Solvent system:
n-propanol-1% aqueous ammonia (2:1).   Nucleosides separated
are inosine and guanosine [12].

(2)   Whatman #1 chromatography paper.   Solvent system:

Isobutyric acid-1M NH$_4$OH-0.2 M sodium ethylenediaminetetra-acetic acid (50:30:0.5 v/v/v).

Spots are located by short-wave ultraviolet light. In (2) the spots are eluted in 70% ethanol, concentrated, and rechromatographed on Whatman #1 paper with water used as the solvent system. The nucleosides separated are desoxyinosine, deoxyguanosine, and deoxyxanthosine. The nucleotides deoxy-adenosine monophosphate and deoxyguanosine monophosphate are also separated [13]. Uric acid riboside may be separated on paper by the method of Krzyanowska and Nemierko [14] as detailed below.

Thin-layer chromatography

(1) Substrate: Cellulose with luminescer (Schleicher and Shull 1446254) [15]. Solvent system: 0.1 Ammonium formate (pH 5.3) ultraviolet light (short-wave) is used to locate the spot, in this case, adenosine. The author verified the identity of the nucleoside by chromatography in 6 solvent systems and by spectra of the eluates at various pH's. Quantitation is by spectrophotometry.

(2) Substrates: (a) Microcrystalline cellulose (Avicel, American Viscose Corp., Marcus Hook, Penn., U.S.A.), (b) DEAE-cellulose. This method [16] was used by Moriuchi et al. [17] with the silkworm, Bombyx mori.

Preparation of the substrates: (a) Avicel:-30 g of commercial grade cellulose are suspended in 500 ml of 5 mM versene and shaken for 10 min, filtered through a Buchner funnel, and washed with distilled water. Repeat the versene treatment; suspend the cellulose in 150 ml of water; and degas

under vacuum. Shake the suspension until a homogeneous slurry
is obtained. (b) DEAE-cellulose:-20 g of DEAE-cellulose are
suspended in 500 ml of 0.5N HCl and mechanically stirred for
30 min at room temperature. The cellulose is collected on a
Buchner funnel and washed with water to neutrality. Resuspend the powder in 500 ml of 5N HCl stir for 30 min, filter
out the cellulose, and wash as before. Then suspend the
cellulose in 200 ml of water, degas under vacuum, and shake
the suspension until a homogeneous slurry is obtained.

With either substrate, the following solvent systems may be
used: (a) Isopropanol-HCl-H_2O (65:16.7:18.3 v/v); (b) Isobutyric acid-H_2O-25% NH_4OH (400:208:0.4 v/v); (c) Isopropanol-
H_2O conc. (28%) NH_4OH (85:15:1.3 v/v); (d) n-butanol-H_2O-for-
mic acid (77:13:10 v/v); (e) saturated $(NH_4)_2SO_4$-1 M sodium
acetate-isopropanol (80:18:2 v/v).

Best results are obtained by using two-dimensional chromatography and employing the solvent systems in Table 1.

Quantitation: The amounts of the nucleosides are estimated
by outlining the spots detected under a short-wave ultraviolet
light and scraping them off into a test tube. HCl (0.1 N) is
added and the resultant slurry is allowed to stand, preferably
overnight. After filtration through a sintered glass micro-
funnel, the solutions are read in a UV spectrophotometer
using differential extinction values (λ max - λ min) [41].
These systems have also been used to obtain effective separation of nucleotides.

Table 1.

R_F Values of nucleosides chromatographed on commercial grade microcrystalline cellulose (Avicel). The time required for migration of all solvents, though variable due to variations in length and temperature, is an average of 2 h.

Nucleoside	Solvent a	Solvent b	Solvent c	Solvent d	Solvent e
Ado	0.23	----	----	0.41	----
de Ado	0.27	0.82	0.37	0.42	0.09
Cyt	0.36	0.65	0.27	----	0.52
de Cyt	0.50	0.65	0.35	----	0.47
Methyl de Cyt	0.59	0.73	0.40	0.42	0.39
Guo	0.22	0.50	0.09	0.26	0.27
de Guo	0.14	0.59	0.17	0.36	0.23
Urd	0.62	----	----	----	----
de Urd	0.77	0.61	0.33	0.47	0.45
Thy	0.88	0.66	0.52	0.55	0.37
Pseudo-Urd	0.42	0.42	0.08	0.24	0.57

Table 1 reproduced from Grippo et al.[16], Biochem. Biophys. Acta 95(1965)4, by permission of Elsevier Publishing Co.

ANALYSIS OF NUCLEOTIDES

Separation procedures for free nucleotides

Concave gradient system: Effective separation of nucleotides of pupae of Tribolium confusum was achieved [5] by the

use of a concave gradient elution system [18] and a mixing
system similar to that of Pontis and Blumson [19].

Materials:

(1) Ion exchange resin AG 1-X4 styrene type, quarternary
ammonium, Calbiochem, Los Angeles, Calif., U.S.A) 200-400 mesh

(2) Glass column (44 x 1.3 cm)

(3) Sintered glass disc for (2)

(4) 1 N NaOH, 1 N HCL

(5) 0.15 N LiCl, 0.16 N LiCl, 0.19 N LiCl

(6) 0.001 N HCl, 0.01 N HCl

Methods:

Preparation of the column: Wash the ion-exchange resin
first with 1N NaOH and then with 1 N HCl. Repeat the cycle;
then wash with water until the pH is above 5. Fill the
column with resin to 32 cm of length.

Elution: Use a concave gradient in a reservoir mixing
chamber. Two glass chambers both 30 cm high but of different
sizes are placed parallel to each other with their bases
joined by a short length of polyethylene tubing. The larger
of two mixing chambers is provided with a magnetic stirring
bar at the bottom. (The authors used chambers 8.6 and 5.5 cm
in diameter.) The following solutions are used:

Reservoir	Mixing Chamber
A. 715 ml of 0.15 N LiCl in 0.001 NHCl	A'. 1685 ml of 0.001 N HCl
	B'. 940 ml of 0.15 N LiCl in 0.001 NHCl
B. 400 ml of 0.16 N LiCl in 0.01 NHCl	C'. 940 ml of 0.16 N LiCl in 0.01 NHCl
C. 400 ml of 0.19 N LiCl in 0.01 NHCl	

The authors note that all 3 eluent solutions are used in the separation of known nucleotide mixtures; however, the B-B' system was omitted for the fractionation of the pupal extract since Tribolium pupae extract contained relatively few nucleotides.

Initially the reservoir and mixing chambers are filled with eluent solutions A and A', respectively, as indicated. The solutions from the reservoir are allowed to flow into the mixing chamber and thereon to the column. When both chambers are nearly exhausted, these are filled with first the eluent solutions B and B' and then with C and C'.

The concentration of the eluent emerging from the column may be calculated as described by Bock and Ling [20]. The fractions may be collected in a fraction collector (5-ml fractions) and the concentration recorded by using a recording spectrophotometer (253.7 nm). For more precise measurements, the fractions may be reestimated by use of a spectrophotometer at 260 nm. Identification of the nucleotides is verified by their elution patterns and absorption spectra.

Column System: This has been used to fractionate nucleotides of Cecropia silk moths [4], and the house fly, Musca domestica [21].

Materials:

(1) Ion exchange resin Dowex 1 x 10 formate and Dowex 50 x 12H

(2) Dowex 1 column 3 - 10 mm x 18-25 cm; Dowex 50 column about twice the volume of the Dowex 1 column

(3) Distilled water

(4) 2N HCOOH, 4N HCOOH-1N $HCOONH_4$

(5) 0.02 N HCl, 0.2 N HCl

Method: On the Dowex 1 column nucleotides are eluted with gradients from water to 2 N HCOOH to 4 N HCOOH-1 N HCOONH$_4$. (Recovery 98 $\pm$ 5.6% S.D.). The Dowex 50 column is used for basic phosphates; these are eluted with water 0.02 N HCl, and 0.2 N HCl successively.

In the formate system at room temperature, no hydrolysis products were produced from an electrophoretically purified sample of ATP. The authors [4] also noted that they used paper chromatography methods for further resolution and identification of the fractions separated by ion-exchange. (See section on Paper Chromatography.)

Separation of cyclic nucleotides

Column System: Cyclic AMP and cyclic GMP were separated from tissues of cricket, Acheta domesticus, prepared as noted within section on sample preparation [8].

Materials:

(1) Glass column 0.7 x 3 cm

(2) Dowex - 1 resin, formate form equilibrated with distilled water, and washed with 10 ml of distilled water.

(3) Formic acid 2M, formic acid 4M

Method: Elute cyclic AMP with 14 ml of 2M formic acid, and cyclic GMP with 18 ml of 4M formic acid. The eluates are lyophilized, and the cyclic nucleotides are dissolved in 0.75-1.50 ml of distilled water. The authors estimate that recovery is 75-80%.

Equilibrium dialysis for cyclic nucleotides [22]:

Materials:

(1) A dialysis chamber with 10 dialysis compartments of
2 x 200 µl

(2) Immune serum (see section "Assay")

(3) Citrate buffer, 0.1 M; pH 6.2

(4) Human serum albumin

(5) Papaverine, 2.5 x 10^{-4}M

(6) ^{125}I 2'-O-succinyl cyclic AMP tyrosine methyl ester
(^{125}ISCAMPTE)

(7) 14C 3',5' cyclic AMP

(8) 2'-O-succinyl cyclic AMP (SCAMP) [23]

(9) Succinic anhydride

(10) Triethylamine

(11) Anti-3',5' cyclic AMP antibody [22]

Methods:

1. Equilibrium dialysis - Routinely a dialysis apparatus
with 10 dialysis compartments of 2 x 200 µl is used [23]. For
higher sensitivity, one with 2 x 20 µl is used. Fill one
chamber with immune serum at a suitable dilution (usually
1/16,000) and the other with ^{125}ISCAMPTE (ca. 10,000 cpm,
2.5 x 10^{-5} mole) plus the unlabeled nucleotide to assay. All
the products are diluted in 0.1 M citrate buffer (pH 6.2) plus
1 g/l human serum albumin x 2.5 x 10^{-4}M papaverine. Equilibrium is obtained in 20 h at 4°C. Count the radioactivity of
10 µl of each compartment. On the immune serum side, it corresponds to the sum, T, of bound and free ^{125}ISCAMPTE and on
the other side to the free ^{125}ISCAMPTE. The ratio of binding
is computed as r = B/T with B = T-F.

Typical curves obtained with carefully calibrated standard
solutions of cyclic AMP and SCAMP are shown in Figure 1.
These will be used for the radioimmunoassay of unknown
samples.

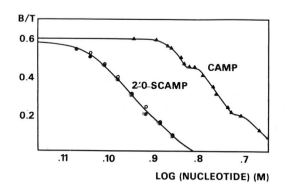

Fig. 1. Displacement curves of [125]ISCAMPTME by 2'-O-SCAMP
and CAMP. The rabbit immune serum is diluted 1/16,000. The
data are plotted as the ratio bound [125]ISCAMPTME / bound +
free [125]ISCAMPTME versus log nucleotide .

2. Succinylation: Add 5 mg of succinic anhydride and
10 µl of triethylamine to 100 µl supernatant. Shake
vigorously for 10 minutes.

Isolation of cyclic nucleotides with thin-layer procedures:

Both cGMP and cAMP of pupae of Hyalophora gloverii have
been isolated on thin-layer plates [24].

Materials:

(1) Thin-layer plates (5 x 20 cm) coated with silica gel
containing an inert binder plus phosphor (type Q1-F, Quantum
Industries, Fairfield, N.J., U.S.A.).

(2) Thin-layer solvent tanks

(3) Solvent systems:

a. Isopropanol-ethylacetate-conc. NH_4OH-water (55:29: 12:4 v/v) [25].

b. Isopropanol-ethylacetate-conc. NH_4OH-water (30:50: 12:18 v/v).

Method: To isolate cAMP, aliquots of a 5% trichloroacetic acid extract of homogenized tissues are spotted on coated TLC plates and run in solvent a. To isolate cGMP, solvent b is used. Markers are run in the usual manner.

The investigators used this method to isolate radioactive nucleotides that were scraped from the plate and counted in a scintillation counter.

Assay procedures

Binding protein procedure for cyclic GMP [8] :

Materials:

(1) Phosphate-EDTA buffer (pH 7.0); 5 mM potassium phosphate containing 2mM EDTA.

(2) Phosphate-EDTA buffer (pH 7.2); 1 M potassium phosphate containing 2 mM EDTA.

(3) Sodium acetate buffer:5 mM (pH 4.0)

(4) EDTA: 4 mM

(5) Acetic acid 1N

(6) Ammonium sulfate

(7) ^3H GMP

Preparation of binding protein: The fat body of Antheraea pernyi pupae is removed, washed three times with Ringer's solution, and drained by gentle swirling on a Buchner funnel (no filter paper). The fresh or frozen tissues are used for isolation of the binding protein [26].

Homogenize the tissue with 3 to 4 volumes of 4 mM EDTA and centrifuge at 27,000 g for 30 min in the cold. After centrifugation, adjust the supernatant solution to pH 4.8 with ice cold 1 N acetic acid with stirring. Allow enzyme solution to stand for 10 min and centrifuge off the precipitate. Adjust the supernatant solution to pH 6.8 with 1 M potassium phosphate buffer, pH 7.2 (contains 2 mM EDTA).

Add solid ammonium sulfate to the supernatant (33 g/100 ml) and stir the solution for 20 min. Collect the precipitate by centrifugation, and dissolve in 6% of the crude extract volume of 5 mM potassium phosphate buffer pH 7.0 (containing 2 mM EDTA). Dialyze the resulting solution overnight against 20 volumes of the same buffer, using two changes of buffer. (Use of more than 20 volumes may cause precipitation.) After dialysis, centrifuge the solution for 30 min at 27,000 g, and discard the supernatant.

Assay conditions: Cyclic GMP is assayed in 0.10 ml total volume containing 5 μmoles of sodium acetate (pH 4.0) 1 pmole of ^{3}H cGMP, and the unknown sample or standard in water. Cool the tubes to 0°C, and carry out all subsequent treatments in an ice bath. Add 420 μg of the binding protein, and allow the tubes to stand for 1.5 h.

Terminate the reaction by adding 10 μl of a 10 mg/μl solution of bovine gamma globulin in phosphate EDTA to the mixture as a carrier [27], followed by 1 ml of saturated ammonium sulfate. The carrier may be routinely added to 8 to 10 tubes before the addition of ammonium sulfate. Mix by hand and then let the tubes stand for about 10 min to allow complete precipitation. Centrifuge at 16,000 g for 10 min. Draw off the supernatant, and resuspend the precipitate in 1 ml of saturated ammonium sulfate and centrifuge again. Dissolve the precipitate in 0.5 ml of the phosphate EDTA buffer and transfer to a scintillation vial. Rinse the tube with 0.5 ml of the same buffer, and add 10 ml of a Triton X-100 based scintillation fluid. In the absence of competing cGMP, 420 µg of silkmoth protein binds about 700 cpm. Blanks with bovine gamma globulin instead of binding protein give about 10 cpm above background.

According to the authors [8], silkmoth binding protein is highly pH dependent, binding is maximal in 0.05 M sodium acetate, pH 4.0. The binding content of a typical protein preparation is 10 mM. Maximal binding is attained at 1.5 h. Only cyclic nucleotides inhibit ^{3}H cGMP binding by silkmoth protein.

For further information on the analysis of cyclic nucleotides, the reader is referred to Gilman and Murad [28]. It is highly probable that other procedures can be readily adapted for use with insect tissue.

Immunological procedure:

Assay:

(1) Assay dilution up to 300 fold.

(2) Add 100 µl of a solution (^{125}ISCAMPTE 20,000 cmp-2g/1-5 x 10^{-4}M, papaverine in 10^{-1}M acetate buffer, pH 6.2) to 100 µl of the solution to assay.

(3) Incubate 15.0 µl of solution (2) to one compartment of the dialysis apparatus, and add 150 µl of the antibody solution to the other compartment. (Antibody 1/16,000-1g/1

albumin 2.5×10^{-4} papaverine in $10^{-1}M$ citrate buffer, pH
6.2.)

(4) Dialyze for 18 h; then draw 1.00 µl from each side
of the apparatus and count. Compute the ratio B/T, and read
concentration of SCAMP from the standard curve.

The authors assert that the sensitivity of this method is
100 times that of the usual radioimmunoassay; $10^{-15}M$ cyclic
AMP is routinely assayed with the minimum detectable being
under 10^{-16} mole.

Paper chromatographic methods:

Niemierko and Krzyzanowska [29], in developing the paper
chromatographic method described, noted that whereas separation methods developed with pure substances worked well, it
was often more difficult to reproduce such results with
extracts of animal tissues, particularly since concentrations
of nucleotides were found in rather low amounts. They were
able, using paper, to apply larger amounts of solutions and
hence larger amounts of the substances being examined. They
applied their method to extracts of the greater wax moth,
Galleria mellonella.

Materials:

(1) Whatman #1 paper, 18 x 18 cm and 18 x 22 cm.

(2) Solvent systems: Solvent A, n-butanol-1% NH_3(60:40);
B, n-butanol-acetone-5% NH_3-acetic acid-water (90:30:20:40
v/v).

(3) 0.4% EDTA

Equipment

For multiple ascending chromatography (MAC):

    Glass jar (20 x 12 x 20 cm high) with polished surface

covered with well-fitting ground glass cover (20 x 12
cm). The size of the paper is 18 x 18 cm. Two perfora-
tions are made in the paper so that it may be hung on a
glass support during MAC (Fig. 2), or to fix it addi-
tionally during continuous ascending chromatography CAC
(Fig. 3).

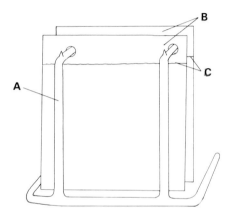

Fig. 2. Arrangement for multiple ascending paper chromatography. A = The glass support; B = the chromatograms; C =
the level to which the solvent moves during each of the runs.

For continuous ascending chromatography (CAC):

A glass jar is used, as before; it is **covered with** 3
tightly adjoining teflon, or gaflon plates (24 x 4.5 x
0.8 cm thick). One avoids condensation on the lower
side of the cover by grounding the gaflon plates as
shown in Fig. 3. When the solvent mixture reaches the
edge of the cover, it evaporates in the air, and as a
result, a further amount of the solvent is forced to

ascend. Towards the end of the procedure, a fairly
straight or wavy line becomes visible at the boundary;
it probably consists of small amounts of impurities
that are still present in the paper and the solvents.

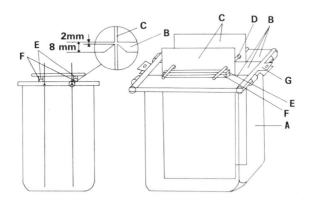

Fig. 3. Arrangement for continuous ascending paper chromatography. On the left, the side view of the jar; in the
middle, a magnified view of the place where the chromatogram
is fixed between the gaflon plates; on the right, a general
view of the whole arrangement. A = The glass jar; B = gaflon
plates used as cover; C = chromatograms; D = glass rods; E and
F = narrow glass plates used for supporting the rods; G =
fasteners.

   This arrangement (Fig. 3) allows for the equilibration of
the paper in the atmosphere of the solvent vapor if necessary.
The sheets of paper may be hung a few millimeters below the
level of the solvent on 2 glass rods supported on two pairs of

narrow glass plates (ca. 8 mm thick). When 2 glass plates, one from each side of the chromatogram (Fig. 3), are taken out, the sheets of paper can be lowered and dipped into the solvent mixture.

Method: Solutions are adjusted to a pH of about 7 and are placed as drops 2.5 cm from the lower edge and 3 cm from each other. These are applied by micropipettes in precise amounts and air dried.

For the MAC procedure, solvent B is used exclusively. In the CAC procedure, the two solvent systems A and B are used successively.

The dried paper is visualized under a short-wave UV lamp. In the event of poor visualization, the dried paper may be dipped or sprayed with a 0.0003% solution of Rhodamine B in absolute ethanol. (The authors [29] assert that this does not affect subsequent spectrophotometric quantitation.) The spots are outlined with a soft lead pencil and are then cut out for subsequent elution and spectrophotometric assay. We have found that elution is easily accomplished (for both nucleotides and nucleosides) by shredding the paper spot into a small test tube (13 x 100 mm) and then adding a measured quantity of a solution of the desired pH, usually a dilute solution of HCl (pH 1) or of NaOH (pH 12). After filtration, the solutions are read in a UV spectrophotometer using differential extinction values [41]. Other solvent systems have been found effective by other researchers. Sevast'yanova and Smolin [30], to confirm identifications of nucleotides found in Antherea pernyi, used:

1) Propanol-ammonia-water (60:30:10)

2) 1 M ammonium acetate buffer (pH 3.8)-ethanol (30:75)

3) 1 M ammonium acetate buffer (pH 7.5)-ethanol (30:15)

4) Methanol-HCl-water (70:20:10)

5) Isobutyric acid-ammonia (pH 3.5)

6) 1% ammonium sulfate-isoproponal (50:100)

Wyatt et al. [31] used several chromatographic solvents to resolve and confirm nucleotide fractions separated by ion-exchange (see section on Ion exchange). These were:

1) Ethyl acetate-acetic acid-water (3:3:1 v/v/v)

2) Methanol-conc.-ammonium hydroxide-water (12:2:3 v/v/v)

3) tert-Butanol-water-picric acid (80 ml:20 ml:2 g)

Enzymatic procedures:

Leenders et al. [3] assayed the amounts of ATP, AMP, and ADP in Drosophilia salivary glands by adapting the methods of Lamprecht and Trautschold [32].

Enzyme preparations:

Hexokinase (E.C. 2.7.1.1.)

Glucose-6-phosphate dehydrogenase (E.C. 1.1.1.49). Ca. 1 mg/ml or about 200 ug protein/ml.

Lactate dehydrogenase. 0.1 mg protein/ml.

Reagents for ATP

1) Triphosphopyridine nucleotide (TPN), ca. 7×10^{-3}M TPN

2) Magnesium chloride, 0.1 M

3) Tris-KOH buffer, 0.1 M

4) Glucose, 0.5 M

5) Potassium carbonate, ca. 5 M K_2CO_3

6) Methyl orange indicator

7) Perchloric acid 40%, ice cold

8) 4% citric acid in Ringer's solution

All solutions should be kept refrigerated at 1-4°C.
Enzyme solutions should be made fresh daily and glucose, and
TPN solutions should be made weekly.

Extraction: After thoroughly washing the tissues with ice
cold Ringer's solution, homogenize in 1 ml 4% citric acid in
Ringer/100 mg of tissue. Then add 0.1 ml of 40% ice cold
perchloric acid, and centrifuge out the protein pellet.
Adjust the supernatant to pH 7.4 with tris-KOH buffer, and
after freezing and removing the excess salt use aliquots for
nucleotide determinations.

ATP assay

Titrate 2 to 4 ml portions with 5M K_2CO_3 with methyl orange
indicator or a titrimeter to pH 7.4. Use a capillary pipette
or equivalent, and stir magnetically throughout the procedure.
Usually 0.12-0.5 ml of carbonate solution is needed. Let the
solution stand in the cold for ca. 10 min, and decant the
solution from the $KCLO_4$ sediment. Use 0.1 or 0.2 ml of this
immediately for spectrophotometric assay.

Spectrophotometric measurements:

Conditions: Wavelength: 366 nm light path: 2 cm. Final
volume for determination of ATP: 5.00 ml. (If the investigator has access to cuvettes with different light paths and
volumes, adjust all volumes accordingly.)

The order of procedure is as follows:

1. Read initial absorbance E_1 for 1-3 min until it remains
constant.

2. Stir into the experimental cuvette the necessary amount
of enzymes.

3. After adding the glucose-6-phosphate dehydrogenase

solution (0.2 ml), read absorbance for 5 min (final value = E_2).

4. Add 0.02 ml G6P-DH to cuvette. After 1 min, read E_3.

$E_3 - E_2 = E_{G6P-DH}$ (caused by addition of G6P-DH)

$E_2 - E_1 = E_{HMP}$

$E_{HMP} - E_{G6P-DH} = \Delta E_{HMP}$ (absorbance change corresponding to the glucose-6-phosphate or hexose monophosphate content).

5. Add 0.40 ml of glucose. Read absorbance E_4 after 30 sec.

6. Add 0.05 ml hexokinase solution to start ATP reaction (usually about 12 min).

7. After 15 min read absorbance E_5.

8. Add 0.05 ml more of hexokinase solution. Read absorbance E_6 after 1 min.

$E_6 - E_5 = E_{HK}$

$E_5 - E_4 = E_{ATP}$

$E_{ATP} - E_{HK} = \Delta E_{ATP}$

Δ_{HMP} and ΔE_{ATP} are used in the calculations.

Calculations:

1. $\dfrac{\Delta E_{ATP} \times V_K \times V_2 \times V_4}{\varepsilon \times d \times V_1 \times V_3 \times V_5}$ = μmoles ATP/g or ml tissue.

2. $\dfrac{\Delta E_{HMP} \times V_K \times V_2 \times V_4}{\varepsilon \times d \times V_1 \times V_3 \times V_5}$ = μl hexose monophosphate/g or ml tissue.

where V_K = final volume in cuvette after last enzyme addition.

V_1 = weight (g) or volume (ml) of tissue.

$V_2 = V_1$ + g(ml) perchloric acid needed for deproteinization.

V_3 = volume of perchloric acid extract before neutralization.

V_4 = V_3 + ml K_2CO_3 needed.

V_5 = volume of deproteinized solution in cuvette.

ε = extinction coefficient for TPHN; E_{366} = 3.3 cm^2/μ mole.

d = light path in cm.

If ATP is to be calculated per g tissue, then both V_1 and V_2 are expressed in grams. If ATP is to be calculated per ml tissue, then V_2 and V_1 are expressed in ml. The weight of the tissue divided by the density of the tissue is V_1(ml). This value is used to calculate V_2(ml). To convert μmoles to μg, multiply by the molecular weight of ATP (507.2) or of G-6-P (260.2).

The authors note that all interference is the result of contamination of the hexokinase of G-6-P-DH with other enzymes. Old enzyme solutions give incorrect results. They also note that inosine triphosphate (ITP) can interfere and that this method gives 10-15% lower values than the phosphoglycerate kinase method [33].

ANALYSIS OF PURINES (OTHER THAN NUCLEIC ACID COMPONENTS)

For many years, all interest in purines in insects was centered on uric acid, which had long been identified as a major component of insect excreta. More recently, research has shown that other purines can be found in insect excreta [34,35,36]. This has resulted in a number of studies of the metabolism of these compounds.

Methods of analysis of the purines generally involve two types of procedure: (1) extraction and enzymatic assay, and (2) extraction, isolation, and spectrophotometric assay. The extraction procedure is usually common to both procedures.

Extraction procedure

Uric acid and other purines such as guanine and xanthine are usually extracted from fecal or tissue homogenates with an alkaline solution such as 0.6% lithium carbonate or 0.1 N sodium hydroxide. For enzymatic assay, the tissues (or feces) are homogenized in a borate buffer (0.05 M; pH 9) or a glycine buffer (0.5 M; pH 9) [37].

In fatty tissues particularly, it is well to remove lipids from the homogenates prior to alkaline extraction. This may be done by homogenizing the tissues with chloroform-methanol (2:1) or ethanol-ether (3:1) and centrifuging off the solvents.

In the author's laboratory, isolation procedures were more successful for purines from both the fecal pellets and the tissues of the boll weevil, Anthonomus grandis [38,36], rather than using enzymatic methods.

Isolation of uric acid, guanine, xanthine, and hypoxanthine:

Grind tissues in a glass homogenizer or feces in a small hammer mill. Extract with chloroform-methanol (2:1) to remove the lipids, and centrifuge off the solvent. Extract the purines with measured quantities of alkaline solution as noted; centrifuge and combine the supernatants. Measured aliquots are applied to the various chromatographic substrates.

Chromatography

Paper chromatography: Whatman #1 filter paper 46 x 57 cm.

Solvent systems for 2-dimensional chromatography:

(a) Sec-butanol saturated with water adjusted to

pH 10 [39].

(b) 65% isopropanol 2NHCl [40].

Solvent system for single dimension analysis: Isobutyric

acid-water-versene-ammonia (100:58.8:16:4.2);

adjust to pH 4.6.

The spots are located by short-wave ultraviolet light.

High voltage electrophoresis. Whatman #4 chromatographic

paper.

Buffer system: pH 1.9 formic-acetic acid buffer

Voltage: 3500 volts

Time: 30 min

Detection is by the methods outlined earlier.

Assay procedures

Uric acid: The uric acid spots are cut out and eluted

overnight in a test tube with 0.1 N NaOH. After filtration,

the eluents are read in a spectrophotometer by using differ-

ential extinction values (λ292 - λ310nm).

Colorimetric determination: Lafont and Pennetier [42] used

the method reported by Carroll et al. [43] to determine uric

acid in Pieris brassicae.

Reagents:

(1) Phosphotungstic acid reagent. Dissolve 40 g of

molybdenum-free sodium tungstate in about 300 ml of water.

Add 32 ml of orthophosphoric acid (850 g/l), and reflux

gently for 4 h. Cool to ambient temperature, and add about 300 ml of water. Then add 32 g of lithium sulfate monohydrate and dilute to 1 liter. Add 8.1 ml of 2N NaOH to 1 liter of the phosphotungstic acid solution. The final pH should be 2.5 ± 0.1, and the solution should be clear and slightly greenish-yellow.

(2) Sodium carbonate solution. Sodium carbonate (anhydrous) 140 g/l.

Procedure: Add 0.2 ml of a crude extract to 2 ml of reagent 1. Let stand 15 min at room temperature and centrifuge. Immediately decant the supernatant into a clean tube, and add 1 ml of reagent 2, and mix. After the mixture stands for 15 min at room temperature, read the absorbance of the sample vs a reagent blank at 700 nm.

The reagent blank should be practically colorless, and the final color is stable for at least an hour.

Uric acid standards are prepared by dissolving uric acid in albumin solution (about 5 g/l).

A relatively simple procedure for the examination of fecal uric acid has been reported [44] using an alkaline neocuproine copper reagent [45].

Reagents:

(1) 40 g sodium carbonate, 16 g glycine, and 0.5 g of cupric sulfate are dissolved in 1 liter of water.

(2) 0.5 g neocuproine hydrochloride is dissolved in 100 ml of water.

The final reagent is a mixture of 10 ml of (1) and 10 ml of (2). It should be discarded when it yellows (within a few hours).

Procedure: An alkaline extract of feces is used. (The author used 0.5% borax.) Aliquots containing 5.20 µg of uric acid are added to 12 mm Coleman cuvettes. (Amounts should be adjusted for cuvettes of other sizes.) One ml of water is added to the cuvettes and in 4-5 min the absorbance is read at 450 nm.

The author [44] has shown that the method is specific for uric acid. It has been used with mosquito feces. Ascorbic acid interferes, and glucose does not react before it reaches several times the uric acid level.

Enzymatic determination: This is the most widely used method and employs the enzyme uricase. Nearly all of the methods reported by various workers are modifications of Kalckar's method [46]. The procedure has the advantage in that it is specific, sensitive, and does not require deproteinization as did many of the older colorimetric methods, although deproteinization may be desirable with certain tissues. The method is based on the fact that uric acid absorbs at 293 nm, but when it is treated with uricase it is converted to allantoin, which does not absorb at that wave length.

The following method [47] is typical and has been used to analyze uric acid in the sawfly, Neodiprion sertifer [48] and with Periplaneta americana [49].

Reagents:

(1) 5 N NaOH

(2) Glycine buffer (0.1 M). Dissolve 7.5 g of glycine in about 800 ml of water. Adjust to pH 9.4 ± 0.1 with 5 N NaOH (about 8 ml); dilute to 1 liter.

(3) Uric acid standard. Dissolve 40.0 mg of uric acid in

100 ml of glycine buffer. This solution is stable when frozen.

   (4) Dissolve 40 mg of dry uricase (1.7.3.3, Worthington
Biochemical Corp., Freehold, N.J.). The activity of the reagent may vary over a 10-fold range and still be satisfactory.
Each new lot of uricase should be assayed for activity.
Pipette 100 µl of the uric acid standard (400 g/ml) into 9.9
ml of the reagent and incubate at room temperature. Measure
absorbance decrease at 292 nm for 10, 20, or 30 min. If rate
of decrease is not within the range of 0.0003 to 0.003 absorbance units/min, adjust by either increasing the amount of
uricase or proportionately diluting the reagent with glycine
buffer.

   Preparation of standard curve: Dilute 5 ml of the uric
acid standard to 200 ml with glycine buffer. Pipette aliquots
of 0, 0.5, 1, 2, 4, 6, 8, and 10 ml into 10-ml volumetric
flask, and make up to mix with glycine buffer. Read absorbances and plot a standard curve of absorbance versus concentration. (Follows Beer's law.)

   Apparatus:

Spectrophotometer with ultraviolet capability.

   Culture tubes 16 x 100 mm: cups about 1 cm in diameter
made by flattening the base of a 1- cm (outside diameter) test
tube and cutting it off about 7 mm from the bottom. A small-
gage wire (25) is bent around the cup, and the ends are
fastened into a hook to hang over the lip of the culture tube.
Capacity of the cup should be great enough to accommodate up
to 200-µl samples.

   Pipettes of varying capacities; for convenience, one that
will repetitively deliver 9.9 ml.

Uric acid determination: Pipette 9.9 ml of uricase reagent into each culture tube. Hook a cup over the side of each tube. Pipette 100 µl of sample, 100 µl of a standard uric acid solution, and 100 µl of distilled water (as a reagent blank) into successive cups. After setting up a series, slip each cup into its culture tube and stopper; shake each tube thoroughly. Then transfer to a silica cuvette and read the initial absorbance (A_o). After marking the meniscus of the remaining solution (use a wax pencil), place the open tubes in a 45° water bath. Then gently bubble oxygen through the solution so it is agitated without spattering. Allow samples to incubate 4 to 16 h, and then make up to mark with distilled water. Read the final absorbance (A_f).

Calculations. Subtract final absorbance from initial absorbance. If necessary, correct the absorbance change by adding any increase shown by the uricase reagent blank. Divide the corrected absorbance and change by the factor k to give the uric acid concentration (µg/ml in the diluted sample). The factor k is the slope of the standard curve (absorbance change $\div$ concentration change). It should have a value near 0.075.

Precautions:

(1) Discard any uricase reagent if the absorbance of its reagent blank is initially greater than 0.200 or if it increases by more than 0.010 upon 45^oC incubation overnight.

(2) Avoid ultraviolet absorbing material (old rubber tubing).

(3) Make sure that glassware has no residual detergent.

(4) Avoid solvent vapors during uricase incubation.

The authors [47] report recoveries of uric acid ranged
from 95 to 104% (mean 100.4%, S.D. 2.6%). The standard
deviations of the absorbances ranged from 0.0008 to 0.0011.
They also indicate that the method is swift and reliable and
that a determination can be completed in 2 to 4 h.

Guanine: After extraction, aliquots of the extract are
spotted on paper and run 2-dimensionally (see section on Paper
chromatography).

After the spots are located under a UV lamp (short-wave),
each is cut out and eluted with 0.1 N HCl. Steeping the spots
in the acid overnight will completely elute the purine. After
filtration (to remove paper particles), the eluates are read
on a spectrophotometer, and quantitative determinations are
made by employing differential extinction values (λ249 - λ290
nm) [41] .

Xanthine: This compound has been found in the feces of the
plum curculio, Conotrachelus nenuphar [36] and may be detected
and quantitated using the same chromatographic methods and
detection systems used for guanine.

Hypoxanthine: First reported to be in the feces of
Melophagus ovinus [50] , and later in a mutant of Drosophila
melanogaster [51] , it was later found in Galleria mellonella
[34] . The latter authors separated the purine from an alkaline extract by using ascending paper chromatography. The
solvent systems found most useful were ethanol-acetic acid-
water (85:5:10), isopropanol-water (10:3), and butanol-
methanol-benzene-water (2:1:1:1). Whatman #1 paper was used.

Quantitation was achieved by cutting out the spots and
eluting them in test tubes. After filtration, the solutions

were read in a spectrophotometer again but using differential extinction values (λ260 - λ290nm).

ACKNOWLEDGEMENTS

The author acknowledges with thanks permission from Analytical Biochemistry, Academic Press; The Journal of Chromatography and Biochimica et Biophysica Acta, Elsevier Scientific Publishing Co., to reproduce the table and figures that appeared in the original articles, respectively [22, 29, 16]. He also wishes to express his gratitude to Carol Vest and Marvel McKee of this laboratory for the typing of the manuscript.

Mention of a proprietary product in this chapter does not constitute an endorsement of this product by the U. S. Department of Agriculture.

REFERENCES

1   S. E. Mayer, J. T. Stull and B. Wastilla, in J. G. Hardman and W. O. O'Malley (Editors), Methods of Enzymology, Vol. 38 (Pt. C), Academic Press, New York and London, 1974, p. 3.

2   O. H. Lowry and J. V. Passeneou, A Flexible System of Enzymatic Analysis, Academic Press, New York and London, 1972, 291 pp.

3   H. J. Leenders, A. Kemp, J. G. Kroninx and J. Rosing, Exp. Cell Res., 86(1974)25.

4   F. G. Garey and G. R. Wyatt, J. Insect Physiol., 9(1963) 317.

5   M. Bouchard, K. D. Chaudhary, J. M. Loiselle and A.
    Lemonde, Can. J. Biochem., 43(1965)1295.

6   K. K. Tsuboi and T. D. Price, Arch. Biochem. Biophys.,
    81(1959)223.

7   M. L. De Reggi and H. L. Cailla, J. Insect Physiol.,
    81(1975)1671.

8   A. M. Fallon and G. R. Wyatt, Analyt. Chem., 63(1975)614.

9   S. J. Berry and W. Firshein, J. Exp. Zool., 166(1967)1.

10 W. Firshein, S. J. Berry and M. Swindlehurst, Biochem.
    Biophys. Acta, 149(1967)190.

11 S. J. Berry, W. Firshein and M. Swindlehurst, Biochem.
    Biophys. Acta, 199(1970)1.

12 H. S. Forrest, S. E. Harris and L. J. Morton, J. Insect
    Physiol., 13(1967)359.

13 L. S. Freeman, M. Swindlehurst and S. J. Berry, Biochem.
    Biophys. Acta, 269(1972)205.

14 M. Krzyanowska and W. Niemierko, Bull. Acad. Polon. Ser.
    Sci., Biol., 18(1970)673.

15 R. Lafont, Experientia 30(1974)998.

16 P. Grippo, M. Iaccarino, M. Rossi and E. Scarano, Biochem.
    Biophys. Acta., 95(1965)1.

17 A. Moriuchi, K. Koga, J. Yamada and S. Akine, J. Insect
    Physiol., 18(1972)1463.

18 P. Lebreton, Bull. Soc. Chem. France, 11-12(1960)2188.

19 H. G. Pontis and N. L. Blumson, Biochem. Biophys. Acta,
    27(1958)618.

20 R. M. Bock and N. S. Ling, Analyt. Chem., 26(1954)1543.

21 J. P. Heslop, Biochem. J., 91(1964)183.

22   H. L. Cailla, M. S. Racine-Weisbuch and M. A. DeLaage,
     Analyt. Biochem., 56(1973)394.

23   H. L. Cailla and M. A. DeLaage, Analyt. Biochem., 48(1972)
     62.

24   S. W. Applebaum and L. I. Gilbert, Dev. Biol., 27(1972)
     165.

25   W. D. Woods and M. B. Waitzman, J. Chromatogr., 47(1970)
     536.

26   J. F. Kuo and P. Greenyard, in J. G. Hardman and B. W.
     O'Malley (Editors), Methods of Enzymology, Vol. 38 (Pt. C)
     Academic Press, New York and London, 1974, p. 329.

27   G. Illiano, G. P. E. Tell, M. I. Siegal and P.
     Cuatrecasas, Proc. Natl. Acad. Sci., U.S.A., 70(1973)
     2443.

28   A. G. Gilman and F. Murad, in J. G. Hardman and B. W.
     O'Malley (Editors), Methods of Enzymology, Vol. 38 (Pt.
     C), Academic Press, New York and London, 1974, p. 49.

29   W. Niemierko and M. Kryzanowska, J. Chromat., 26(1967)424.

30   G. A. Sevast'yanova and A. N. Smolin, Biophysik 32(1967)
     452. (In English.)

31   G. R. Wyatt, R. Kropf and F. G. Carey, J. Insect Physiol.,
     9(1963)137.

32   W. Lamprecht and I. Trautschold, in H. U. Bergmeyer
     (Editor), Methods of Enzymatic Analysis, Academic Press,
     New York and London, 1963, p. 543.

33   H. Adams, in H. U. Bergmeyer (Editor), Methods of Enzyma-
     tic Analysis, Academic Press, New York and London, 1963,
     p. 539.

34   J. L. Nation and R. L. Patton.   J. Insect Physiol.,
     5(1961)299.

35   J. L. Nation, J. Insect Physiol., 9(1963)195.

36   N. Mitlin and D. H. Vickers, Nature 203(1964)1403.

37   P. Plesner and H. M. Kalckar, in D. Glick (Editor),
     Methods of Biochemical Analysis, Interscience Publishers,
     New York, 1954, p. 97.

38   N. Mitlin, D. H. Vickers and R. T. Gast, Ann. Entomol.
     Soc. Am., 57(1964)757.

39   K. Fink, R. E. Cline and R. M. Fink, Analyt. Chem., 35
     (1963)389.

40   G. R. Wyatt, Biochem. J., 48(1951)581.

41   E. Vischer and E. Chargaff, J. Biol. Chem., 14(1948)121.

42   R. Lafont and J. L. Pennetier, J. Insect Physiol., 21
     (1975)1323.

43   J. J. Carroll, H. Coburn, R. Douglass and A. L. Babson,
     Clin. Chem., 17(1971)17.

44   E. Van Handel, Biochem. Med., 12(1975)92.

45   S. Dygert, L. H. Li, D. Florida and J. A. Thomas, Analyt.
     Biochem., 13(1965)374.

46   H. M. Kalckar, J. Biol. Chem., 167(1947)249.

47   C. A. Dubbs, F. W. Davis and W. S. Adams, J. Biol. Chem.,
     218(1956)497.

48   W. H. Fogal and M. J. Kwain, J. Insect Physiol., 20(1974)
     1287.

49   D. E. Mullins and D. G. Cochran, J. Exp. Biol., 16(1974)
     557.

50   W. A. Nelson, Nature, 182(1958)115.

51   H. K. Mitchell and E. Glassman, Science, 129(1950)268.

CHAPTER 2

THE BIOCHEMICAL ANALYSIS OF INSECT DNA

BETTY JANE MILLS, WALTER MASTROPAOLO and CALVIN A. LANG

Department of Biochemistry and the Biomedical Aging Research Program

University of Louisville School of Medicine, Louisville, KY 40201, U.S.A.

CONTENTS

INTRODUCTION

Insects have played an important role in DNA research involving a variety of investigations including aspects of molecular genetics, developmental biology, biochemistry, and cell biology. Indeed one of the most important, early contributions to our understanding of DNA structure was the development and application of a chromatographic method to determine the composition of locust DNA and other DNA's by the insect biochemist, G. R. Wyatt (1). His accurate results on the purine and pyrimidine composition confirmed Chargaff's rules on the pairing of bases. This concept of complementary bases was an essential factor in the development of the double helix model of Watson and Crick (2).

There are a number of biological features found in insects which make them appropriate organisms for the study of DNA. They bridge the gap between microbial and mammalian systems in several ways. Insects resemble mammals since they are multi-cellular, eucaryotic organisms. On the other hand, insects can be readily cultured under controlled conditions like microorganisms. Further, insects have an additional advantage, for they can be cultured singly or in masses. Another unique feature that is not present in either microbial or mammalian systems are the discrete developmental stages of certain insects which allows the correlation of molecular and morphological events with specific developmental phenomena that are observed at the cellular, organ, and organismic levels.

There has been an increasing amount of work reported in the literature on the biochemistry of insect DNA's which indicates an awareness of the advantages of the insect system and wider appreciation of insect models.

Our experience with insect DNA has been almost exclusively with the yellow fever mosquito, Aedes aegypti, which we use as an experimental model for various studies of the biochemistry of growth and aging. For this reason, the major framework of this review will consist of our experience with mosquito DNA's interspersed with results of others and will indicate some of the special problems we encountered.

Our entry into DNA studies stemmed from a peculiar finding during our initial studies of growth and aging in the mosquito. The original and simple purpose was to determine protein, DNA and RNA concentrations throughout its life span, to define the developmental stages in biochemical terms and to use this

information as a reference basis for expressing results in later experiments (3).
During the course of this study subcellular fractions were analysed and a low
molecular weight DNA was found in high concentrations in the high-speed supernatant fractions of homogenates of larvae. but was less than 10% of that value in
adult mosquitoes (Table 1). Subsequent results demonstrated that the concentration of supernatant DNA was correlated only with the period of rapid larval
growth. This unexpected finding prompted us to examine further this unique
DNA, which we called sDNA because of its occurrence in the soluble fraction. in
contrast to bulk DNA, which was termed pDNA because it pelleted with the
particulate fraction (4).

Developmental Stage	Fraction	% of Homogenate	
		Protein	DNA
Larval	Homogenate	100	100
	Nuclear	24	39
	Mitochondrial	15	13
	Microsomal	12	3.1
	Soluble	56	31
Adult	Homogenate	100	100
	Nuclear	46	74
	Mitochondrial	9.2	7.6
	Microsomal	11	2.2
	Soluble	29	1.5

Table 1. Subcellular Distribution of DNA in the Mosquito

Each value is the average of three or four experiments. The range of recovery
was 86-107%.

With the discovery of sDNA we considered carefully the merits of continuing
in this new direction because of the unknown difficulties of DNA biochemistry.
However. the decision was made to pursue the study of sDNA. and it became
apparent that the most important question was "What is the role or function of
sDNA?" To this end. we took two different approaches. The first was an in vivo
aspect to determine the possible relationship of sDNA to the biosynthesis or
degradation of bulk DNA. The second was an in vitro approach to isolate and
characterize sDNA and pDNA and to compare their physical-chemical properties.
The results indicated that sDNA plays a role in replication and that isolated

sDNA has the same composition as pDNA but is smaller in size. This led to our
present studies to purify and characterize DNA polymerase and related enzymes
to elucidate the molecular mechanism of sDNA synthesis.

The need for this review was questioned originally, for there is a surfeit of
DNA reviews in the literature. However, upon reflection of our early experience
and problems, we realized that the best reviews and articles on DNA analysis
and characterization are scattered and difficult to identify. Further, they deal
primarily with microbial, viral and mammalian systems and thus are not
directly applicable to insects. For this reason we decided to summarize our experience, to present some basic rationale and strategy that facilitate the choice
of methods to fit specific needs, and to indicate some useful literature references.

The information in this paper was derived from three sources, namely,
literature references, our own experience, and probably most important, the
advice of a number of DNA investigators on a variety of problems. We have
restricted the discussion to procedures that have been useful in our laboratory
and therefore, make no claim that our coverage is encyclopaedic.

Our aim is to provide basic information and references for the entomologist
or biologist who is a novice to the DNA field. Hence, some of the information
may be too simple for those who already are acquainted with molecular biology
techniques. It is also likely that our methods may be inappropriate for other insect systems. However, we feel that some of the aspects may be applicable,
and if the reader is helped in finding a useful reference or procedure, we will be
satisfied.

                                    EXPERIMENTAL STRATEGY

The Question - Why are you planning to study DNA? This question regarding the
rationale and strategy of any research is the most important and fundamental
initial consideration. And it is too often overlooked or unclear.

A classical and common reason for determining DNA content in insect tissues is to use this measurement as an index of growth and developmental status
since it is a crude estimate of cell number. For this reason, DNA content is
used as a basis for expressing other biochemical measurements such as protein,
RNA, or enzyme activity levels. Although DNA content may be a misleading

index of cell number because of polyploid and polytene cells, it still may be considered the only biochemical estimate available. In many instances the protein content, which is easier to determine, can serve equally well.

The study of various functions of DNA is probably the major reason for DNA analyses today. These areas include DNA biosyntheses such as replication and repair and the regulation of these processes by hormonal or nutritional modifications. Other studies include cell and molecular biological aspects such as transcription, gene amplification, and the role of repetitive and unique sequences. The evolutionary and species-specific relationships of insect DNA's have been explored very little and are potentially fruitful problems for modern insect taxonomists.

Source of DNA - Ideally, an insect is selected as the experimental organism because it is uniquely appropriate for the problem. In practice, however, most laboratory investigators usually restrict their research to a particular insect order or genus with which they have had previous experience. In any case, the biological characterization of the insect species should be considered in regard to its genetic homogeneity, developmental stages and nutritional history.

Fundamental to most studies is the assurance that the DNA examined is derived from the organism studied, rather than from associated microflora or from intracellular symbionts. The investigator should be aware of this possible contamination and the effects it may have on different analyses. For example, in radiometric studies comparing DNA synthesis in axenic versus standard-cultured mosquito larvae, we found distinct quantitative differences between the two groups. Although axenic culture methods are not feasible for most insect species, their use should be considered and encouraged.

For some problems it may also be important to consider the source of the DNA in regard to body region and tissue distribution or subcellular localization. The relative concentration of DNA may differ according to the degree of cellularity, polyploidy or polyteny in the tissues. For example, in the adult mosquito the head contains about 20% of the total DNA content in the whole organism, but only 7% of the weight and protein due to the concentration of cells in the eyes (Table 2). The high degree of polyteny in Dipteran salivary glands has been exploited to good advantage in both biochemical and cytological investigations.

Developmental Stage	Region	% of Homogenate		
		Weight	Protein	DNA
Larval	Head	14	11	21
	Thorax	21	26	33
	Abdomen	65	63	46
Adult	Head	7	6.5	21
	Thorax	57	66	39
	Abdomen	36	28	39

Table 2. Distribution of DNA in Different Regions of the Mosquito

As an investigative tool subcellular fractionation by differential centrifugation has increased the specificity of many analyses and allowed greater differentiation and localization of cellular activities. The obvious application of this technique is in the isolation of nuclear and mitochondrial DNA's, but as stated earlier, our discovery of sDNA in the mosquito was based on distribution studies (Table 1). A more widespread use of this procedure in DNA studies may reveal interesting changes in other insects, particularly as they relate to developmental processes.

A major advantage in the use of holometabolous insect models is their distinct developmental pattern, and for a number of species hormone titers during development have been determined. Although the direct effects of juvenile hormone and ecdysone on DNA metabolism have not been elucidated, most investigators are aware of the possible implications. Less obvious to the novice may be other DNA differences related to developmental status which involve not only total amount, but also size. synthetic rate, and doubling time. Although different causal factors have been suggested. several investigators have noted the difficulty in obtaining DNA of high molecular weight from insects at certain developmental stages (5,6). Also the rates of DNA synthesis may differ not only in relation to overall larval growth. but also according to intramolt patterns (7,8).

In addition, differential rates of DNA synthesis have been noted for a variety of tissues in the same insect at the same development stage (9,10). Similar changes in the duration of the cell cycle or in the deoxyribonucleotide pools may be important considerations for certain studies.

Choice of Method - The major problem and strategy after identifying the kind of information required is to select and develop a procedure to fractionate tissue
samples appropriately for the subsequent biochemical analysis. The distinction in
choice of fractionation methods depends on whether the experiment requires quantitative recovery and thus an accounting of the DNA or its metabolites in all fractions,
or requires isolated DNA.

Studies at the tissue and cellular levels require measurement of all the DNA
present. and purity is relatively unimportant. Typical experiments have included
the DNA content of tissues as a biochemical index of cell number. the biosynthesis
of DNA. and the determination of age profiles in developmental studies. The
critical factor is that the method is quantitative and specific for DNA. and this
must be ascertained and not assumed. It should be emphasized that quantitative
procedures are also used to monitor the purification and isolation of DNA because
in our experience and that of others. preferential losses of nascent DNA occur in
some steps (11. 12. 13. 14). Studies at the molecular level. however. require
highly purified or isolated native DNA which cannot be obtained without some loss
in yield.

Analytical Rigor - Rigorous analytical methodology requires that several essential
conditions are met. Since classical, analytical chemistry courses have been
dropped from many chemistry and biology curricula, these conditions are unknown
or overlooked with the result that too many biochemical measurements appearing
in the literature are questionable. This is a common criticism in the evaluation
of research grants and manuscripts. and for this additional reason we have included a brief outline of these conditions: (a) Proportionality between sample size
and optical density (or other measurement). This is Beer's Law of photometric
analyses or more generally a dose-response curve. Determination of the linear.
proportional region of a standard curve is essential to determine the sensitivity
and limits of concentration which is information required for subsequent dilutions
of samples prior to analysis. We routinely run a standard curve with each experiment to correct for daily variation and also run several dilution levels of each
sample to verify that the sample measurements are specific for the compound in
question and are not due to interfering substances. (b) Recovery of known
amounts of a standard or purified DNA that are added to the unknown sample is
another important procedure to evaluate possible losses during tissue processing

or e.traction techniques as well as the analytical procedure. Recovery should be determined for each of the different tissues or developmental stages being investigated. The use of radioactive DNA facilitates this procedure, and in addition, may indicate new forms of DNA that would be missed. (e) Finally, adequate statistical sampling, design, and analysis of data are essential. One should realize that different samples from different batches or populations are desirable for replication, and not merely aliquots or portions of a single sample. Also note that for the determination of any curve, whether a standard curve or one describing changes during insect development, it is better to sample a wide range of points rather than multiple analyses of a few points. This is especially pertinent to developmental patterns in which important and interesting changes occur that could be missed if only a few points are studied. In practice, a maximum precision level of 10% (ratio of SEM/mean x 100) is often sufficient, for this level is usually less than the biological variation among organisms, which often is the overall limiting variation of such studies.

Valid DNA analyses are within the realm and capability of most entomologists and biologists. The reader should be encouraged by the fact that the processing procedures and biochemical determinations are simple and feasible in any laboratory which has the basic equipment of a refrigerated centrifuge, ultraviolet spectrophotometer, simple chromatography apparatus, and beta scintillation counting equipment. Moreover, the relatively inexpensive availability of various isolated DNA's, the constituent purine and pyrimidine bases and nucleotides, and specific enzymes, such as DNAses, RNAses, and proteinases, now make DNA research simple and far removed from its infantile period.

At the risk of belaboring the obvious, we have included a number of considerations which are fundamental to all experiments and not only to DNA research. Our reason is that these points are often overlooked in DNA experimentation with the result that information is gathered which does not answer the specific objectives of the experiment or which has limited relevance to any question. Perhaps we have become too indoctrinated with the original concept that DNA is a single, metabolically stable entity that exists only as a very high molecular weight polymer. In recent times our knowledge has been extended, and it is realized that DNA exists in the cell and in the test tube in a variety of molecular forms with different physico-chemical characteristics, and these forms are part of a dynamic, metabolic system.

BIOLOGICAL BACKGROUND AND TREATMENT

Genetic Homogeneity - The problem of genetic variation among species is well-known to biologists, but it should be noted that strain differences can also occur and should be considered. To minimize this, we use an inbred colony of Aedes aegypti that has been maintained in our laboratory for about 16 years and over 200 generations. Periodically we compare results with our earlier data and with strains obtained from other laboratories. Also our larval rearing room is remote from the adult cages to obviate contamination from other strains and species.

Culture Conditions - Insect research utilizes field-collected as well as laboratory-colonized insects. Although genetic homogeneity and environmental history are better controlled under laboratory conditions, the colonization of many species of insects is either impossible or unfeasible. However, several considerations that apply mainly to laboratory colonies may also be applicable to collected specimens that are maintained only briefly in the laboratory.

Any environmental variable that affects development or aging must alter the metabolism of the insect and thus can potentially affect DNA metabolism or function. Two major environmental factors are temperature and nutrition. It is well known that the rearing and maintenance temperatures affect metabolic and developmental rates since insects are poikilothermic organisms. Within about a 10° C temperature range (22-32° C) one can manipulate the metabolic rate of the mosquito with no harmful effects. However, this range may be different for other insects, and one should beware of temperature stress phenomena which will complicate the interpretation of results. We found that lower rearing temperatures increased the development and survival times and resulted in marked increases in all parameters of body composition of mosquitoes (15). Temperature-induced changes such as these may be related to DNA metabolism.

The other important environmental variable is the nutritional history of the organism. The use of a defined diet or culture medium is highly desirable in order to achieve strict control of the nutrient intake. The common practice of rearing mosquito larvae on a bacterial infusion using animal chow pellets may not be sufficiently defined or controlled for many biochemical or developmental studies, especially if radiolabelled DNA precursors are used. We developed an axenic culture method for mosquito larvae which has enhanced the specificity of our experiments both by the use of a defined medium and by the exclusion of the intestinal microflora which are normal dietary constituents for various Diptera (16). Larvae cultured by this method have the same rates of growth, pupation and adult emergence as standard-reared larvae. Intracellular symbionts which have been a problem in nutritional studies of the cockroach are absent in our mosquito strain.

Developmental Stages - The developmental stage of the insect can be an important variable. Biological differences between larvae, pupae and adults are well known, but the use of more precise ages is desirable, for in mosquitoes and other insects, there are biochemical and DNA changes which occur specifically after larval and larval-pupal molts, after adult ecdysis, and during early and late adult ages. The precise identification of discrete stages as evidenced in holometabolous insects should be exploited to achieve samples which are synchronous in development. Synchronization of the yellow-fever mosquito can be readily achieved at egg-hatching, for eggs of this species can be stored and hatched within a few minutes. Also the mosquito can be synchronized at the time of each larval molt, pupation, or adult emergence. In the case of diapausing Lepidoptera further development is triggered by a temperature shift. This feature of synchrony permits the magnification of transient changes during development which may otherwise not be detected in an asynchronous population.

Experimental Treatment - The administration of chemical compounds is an essential aspect of basic and applied insect research. These compounds include hormones, specific metabolic inhibitors, nutrient substances, and toxicological agents. Toxicity levels for all compounds should be determined at the beginning, for there have been a number of confusing reports with DNA and RNA inhibitors in other systems that were due to toxic and lethal dose effects. Irradiation treatment is another important experimental manipulation that can be useful in DNA studies.

THE QUANTITATIVE DETERMINATION OF DNA

The major problem in the quantitative determination of DNA in tissues is the removal of interfering substances such as nucleotides and RNA which give false-positive results. Therefore for the determination of total DNA it is essential to process and fractionate samples so that DNA is separated from these interfering substances with minimal loss of DNA. Although the selection of a tissue-processing method depends upon the subsequent analysis. basic general requirements to minimize DNAase degradation include the use of fresh tissues, cold conditions. and the addition of inhibitors. The possibility of DNA degradation by nucleases is a consideration of prime importance. particularly if the size of the DNA or the acid-soluble pool is a key measurement. The nuclease activity of tissue samples may be evaluated by measuring their DNA content after incubation for different periods of time. Quantitative recovery of the DNA or of known amounts of added DNA is indicative of the absence of high levels of nuclease activity. To inhibit enzymatic degradation of DNA, chelating agents, detergents and protein denaturants can be added in the initial steps of processing.

Homogenization and Subcellular Fractionation – In any case, the first step is usually a procedure to mince or grind up the tissue and disrupt cell boundaries to produce cell-free homogenates. We have most frequently utilized an all-glass Ten Broeck homogenizer with a motor attachment to drive it at about 1000 rpm. Microscopic examination of the cell-free preparations of larval and adult mosquito samples produced in this way showed considerable fragmentation of nuclei and mitochondria. To obtain better nuclear preparations, we have used a Dounce homogenizer or Teflon-glass types, and these may be more satisfactory for other systems.

The high activity of polyphenol oxidase (tyrosinase) in larval samples produces a dark pigment in homogenates which may interfere with colorimetric measurements. A few crystals of phenylthiourea placed in the homogenizer before adding the sample will prevent this darkening. The addition of crystals rather than a solution is necessary to obtain an effective concentration of this relatively water-insoluble compound.

A mechanical dissecting system has been described for the large-scale isolation of various organs such as salivary glands, imaginal discs and testes, from <u>Drosophila</u> <u>melanogaster</u> (17). The procedures which depend on differences in shape, size and density of the different organs have been used to study nuclear proteins and nucleic acids.

Subcellular fractions of tissue homogenates can be obtained by the classical differential centrifugation technique of Schneider and Hogeboom (18). To preserve nuclei or other organelles during this procedure, a medium of 0.25 M sucrose or 0.15 M (0.9%) NaCl is suggested, for these are iso-osmotic solutions for the mosquito as well as mammalian tissues for which they were devised originally. The addition of divalent cations (Ca^{2+} or Mg^{2+}) in the medium may help to preserve nuclear integrity. A number of modifications of the Schneider-Hogeboom method are now in use, and current literature should be consulted for details. A table showing the effects of different reagents on nuclei is found in a paper by Howell (19).

Total DNA is released from nuclei by the use of detergents such as sodium dodecyl sulfate (SDS) or sodium para-aminosalicylate. In addition, DNA can be dissociated from protein by homogenization of tissues in concentrated solutions of urea, sodium perchlorate ($NaClO_4$) or sodium thiocyanate (NaSCN). Methods using concentrated salt solutions have the additional advantage of inhibiting enzymic degradation in the initial step.

<u>Separation</u> <u>from</u> <u>Interferences</u> - Subcellular fractionation is not adequate to separate DNA from DNA precursors which give false-positive results in colorimetric assays. Moreover, determination of DNA in the nuclear fraction may not be a measure of <u>total</u> DNA because significant amounts of DNA such as sDNA or mitochondrial DNA do not sediment with nuclei.

DNA can also be separated from other cellular materials on the basis of its large size, density, or solubility characteristics. In addition, its stability in the presence of alkali or RNAase permits separation from RNA.

A generally useful procedure to separate DNA from low molecular weight substances is to precipitate DNA with other macromolecules by the addition of perchloric acid (PCA) or trichloroacetic acid (TCA) or by the addition of ethanol to a concentration of 75% or greater. An alternative method is the differential precipitation of nucleic acids and proteins from aqueous solutions by adding a mixture of ethanol and perchloric acid (20). In one step, 95-100% of the nucleic acid was removed from detergent extracts of cultured plant cells, and this

procedure may be useful in removing interfering substances prior to the chemical determination of nucleic acids. Other separation procedures include gel
filtration, equilibrium centrifugation, and hydroxyapatite chromatography.

Photometric Determination - The colorimetric determination of DNA is both
sensitive and specific and requires minimal purification of the DNA. Although
UV absorbance at 260 nm may also be used for quantitation, it requires highly
purified DNA which can only be obtained with some loss in yield. However, the
colorimetric procedures are not without pitfalls and must be carefully evaluated
for each kind of tissue analyzed. Munro and Fleck (21) have reviewed in detail
the various methods, and several which have appeared in the literature since
1966 will be discussed later in this section.

The most widely used method for DNA determination is the Burton modification of the Dische diphenylamine (DPA) reaction (22). DPA reacts specifically
with the deoxyribose moiety of purine nucleotides to form a colored derivative
and thus obviates interference from RNA. The Burton method was used by this
laboratory for the determination of DNA in tissues of post-embryonic Aedes
aegypti (3). The basic steps of this procedure were as follows: Nucleic acids
were precipitated from an homogenate or subcellular fraction by addition of dilute
perchloric acid to a final concentration of 0.25 M, and the mixture was stirred
for 5 minutes. After centrifugation the acid-insoluble fraction was washed with
dilute $HClO_4$, and then extracted twice with 95% (v/v) ethanol to remove other
interferences. Nucleic acids were extracted by two treatments of 1.0 N $HClO_4$
for 30 minutes at 80° . A portion (0.3 ml) of this extract was combined with
DPA reagent (0.7 ml) and incubated at room temperature overnight. [DPA
Reagent - Dissolve 1.59 g of DPA in 100 ml glacial acetic acid containing 1.50 ml
concentrated H_2SO_4. On day of assay, 0.1 ml of 1/20 dilution of acetaldehyde
is added to 10 ml of DPA solution.] The absorbance at 600 nm of each sample
was determined and compared with a standard curve constructed using commercial calf thymus or salmon sperm DNA treated similarly.

Church and Robertson evaluated DNA determination methods for Drosophila
tissues and their method was similar to our procedure for Aedes (23). The
principal differences were tissue homogenization in 95% ethanol and extraction
of interferences with chloroform-methanol. Their developmental profile of DNA
content for Drosophila larvae paralleled our results with Aedes.

The necessity for careful evaluation of methods for each species and tissue
was illustrated by the work of Linzen and Wyatt in silkworm tissue (24). They
reported that dialysis of fat body preparations facilitates extraction of DNA with
hot acid and also obviates interference due to the high uric acid content. Dialysis
was not necessary in wing tissue from the same organism. Since the fat body is
small and the uric acid content low in the mosquito, the dialysis step may be
eliminated. Measurements of DNA content by the Burton method also have been
made in Tribolium confusum (25) and Calliphora (26).

More recently Richards et al. reported modifications that simplified the
DPA procedure, increased sensitivity by 30% and increased the stability of the
reagents, but their application to insect tissues needs evaluation (27). Additional
methods for DNA determination have appeared in the literature, and although
they have not yet been validated for insects, they offer certain advantages which
merit consideration. For example, a method in which p-nitrophenylhydrazene
reacts with the deoxyribose of DNA to yield a purple chromophore, was originally
described by Webb and Levy (28) and has since been modified by Martin et al.
(29). This procedure is slightly more sensitive and specific than the diphenylamine method and should be considered in cases where interfering substances
present a particular problem. Another advantage is the combination of chromophore development with hot acid extraction which minimizes the possibility of
low values due to destruction of deoxyribose by hot acid. Deoxyribose
destruction is a significant problem in some systems when using DPA and
requires careful evaluation of acid concentration and temperature as well as
correction for losses (30).

Extremely sensitive fluorometric techniques for DNA determination should
be considered where the availability of tissue is limited. Most colorimetric
procedures require a minimum of 10 μg of DNA or 50-100 mg of tissue. In
contrast, Kissane and Roberts describe a microfluorometric technique that requires a minimum of 2.5 ng of DNA and 10 μg of tissue (31). This technique is
based on the reaction of 3,5 diaminobenzoic acid with deoxyribose to yield a
fluorescent product. The reaction is highly specific for deoxyribose as opposed
to ribose and, therefore, separation of DNA from RNA is not required. Another
advantage is that a separate extraction step with hot acid is not necessary, thereby eliminating the possible destruction of deoxyribose discussed earlier. Other

fluorometric procedures for DNA analysis in tissues have been described (32, 33).

The older literature prior to 1956 contains accounts of DNA determination by measurement of phosphorous content, a procedure which may lead to considerable error due to the presence of RNA if purity of the DNA is not assured. The obvious lack of specificity of phosphorous determination would require more extensive purification of DNA with concomitant risk of loss. We, therefore, do not recommend this procedure for determination of DNA content in tissues.

THE RADIOMETRIC DETERMINATION OF DNA BIOSYNTHESIS

Biosynthesis - The interpretation of in vivo synthesis of DNA and other macro-molecules is sometimes confusing, for the distinction between the rate and the extent of synthesis is often not made or appreciated. Synthetic rates are a measurement of capacity and are usually determinations of the initial rates of incorporation of a precursor over a period of minutes or hours. In contrast, the extent, or the steady-state level, is a measurement of the total DNA content over a period of days or weeks.

There is no simple relationship between the rate and extent of DNA synthesis, for the measurement is complicated by cell turnover, DNAase action and DNA repair processes. Hence, it is difficult to compare initial rates with long-term, tissue levels. Moreover, in vivo studies have the additional problem of changes in the size or the specific radioactivity of the nucleotide pool as measured in the acid-soluble fraction, which can alter the measured rate.

In spite of these considerations, the in vivo labelling of DNA by the administration of radioactive precursors and density-labelled analogs is a powerful, sensitive, and quantitative tool for the investigation of DNA biosynthesis. There are many applications of the radiolabelling technique, and most procedures measure the radioactivity of a specific DNA precursor which is incorporated into an acid-insoluble fraction. Digestion of the radioactive, insoluble material by DNAase to acid-soluble products demonstrates that the incorporated radioactivity is in DNA. The radiometric procedure lends itself to precise quantification using small amounts of material, and recovery of radioactivity in the fractions relative to that of the homogenate can be determined in all cases.

The procedure which we used for the preparation of acid-insoluble, sDNA and pDNA fractions for radioactive counting is outlined in Fig. 1.

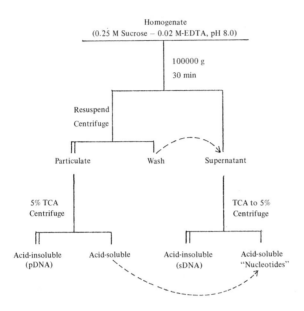

Fig. 1 Sample preparation scheme

Sample Preparation - Cold-inactivated, radiolabelled larvae or adults were homogenized in 0.25 M-sucrose/0.02 M-EDTA, pH 8.0, by using an all-glass Ten Broeck homogenizer. A portion of the homogenate (5-20% w/v) was centrifuged 30 min at 100,000 g at 4° C in a SW 50.1 swinging-bucket rotor in a Beckman model L ultracentrifuge. The supernatant was removed, and the pellet was resuspended in homogenizing medium and recentrifuged under the same conditions. The washing was combined with the first supernatant. The fractions were acid-precipitated by the addition of TCA to a final concentration of 5% (w/v) and centrifuged at 5500 rev./min for 20 min to separate the acid-insoluble and acid-soluble materials. On the basis of the results of Kao et al. (34) the radioactive, acid-precipitable material of the supernatant fraction was defined as "sDNA" and that of the particulate fraction as "pDNA." These fractions were solubilized in 1 ml of Protosol at 55° C, and 10 ml of scintillation fluid were then added. Protosol was also added to portions of the homogenate and to acid-soluble fractions, followed by incubation and the addition of scintillation fluid. All procedures before Protosol treatment of the samples were carried out at 0-4° C. The samples were analyzed for radioactivity in a liquid scintillation counter by using channels-ratio settings.

Although the acid precipitates can be collected on glass-fiber disks, rinsed successively with TCA, ethanol, and ether and then counted for radioactivity, there were still considerable differences between the fractions in the degree of quench which must be corrected. Therefore, we found the single-phase system using a solubilizer like Protosol, coupled with quench-correction calculations, more sensitive and reproducible than the two-phase disk method.

We have utilized continuous or pulse-labelling procedures singly, in combination, or followed by a "chase" with a non-radioactive precursor to determine changes in DNA content and synthetic rates throughout the mosquito life span and to monitor the fate of newly-synthesized DNA (34, 35).

Continuous Labelling - The changes in DNA content throughout the life span were quantified from the incorporated radioactivity of mosquitoes grown in culture medium supplemented with $[^{14}C]$thymidine from the time of hatching until pupation. The importance of such a procedure to determine the steady-state DNA content was pointed out by Selman and Kafatos (36) who found that thymidine administered to silkworm pupae by pulse-labelling disappeared from the nucleotide pool in 2 to 4 h, and thus pulse-labelling is applicable only for determining short-term, synthetic rates. For continuous administration of the precursor pool, we added about 60 newly-hatched mosquito larvae to 250 ml of defined axenic culture medium containing 1 μCi of $[^{14}C]$thymidine. They were grown in this radioactive medium until removal for sampling, or until pupation, when the pupae were transferred to distilled water from which adult mosquitoes emerged. Larval development, pupation, and adult-emergence rates were the same as for mosquitoes cultured in non-radioactive medium. Although survival curves were not established for the radioactive populations, we believe they were normal, for mosquitoes 40 days of age or older were easily obtained, and no apparent deleterious alterations were observed in the larvae or adults. A similar method was used by Chaudhary and Lemonde who grew Tribolium confusum larvae in wheat flour containing $[^{14}C]$thymidine (37).

From the age profile obtained by this method we found that mosquito larvae showed exponential increases in DNA content which paralleled the increase in weight for the first 4 days (Fig. 2). The doubling time for DNA during the first 4 days was 10.5 h according to a least-squares calculation. Although the pDNA component was approximately ten times that of sDNA, the two profiles were similar. There was also a close correlation of the sDNA and pDNA profile with that of DNA content determined previously by a diphenylamine colorimetric procedure (Fig. 3). A marked lag in both profiles occurred between days 4 and 5.

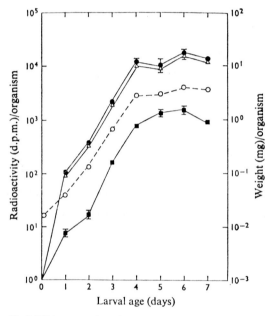

Fig. 2 *DNA contents throughout the larval stage of the mosquito*

Axenic larvae continuously-labelled with [^{14}C] dT were sampled at 24h intervals. Pupation occurred between 6 and 8 days. Fresh weight and content of radioactive DNA species are plotted semi-logarithmically. Each point with vertical bar represents the mean ± S.E.M. of eight to eleven samples from three or four different hatches. Bars are omitted if the S.E.M. is less than the size of the point. From 68 samples there was an overall recovery of 99.3 ± 0.87% (radioactivity of all fractions/homogenate radioactivity × 100). ●, sDNA + pDNA; △, pDNA; ■, sDNA; ○, fresh weight.

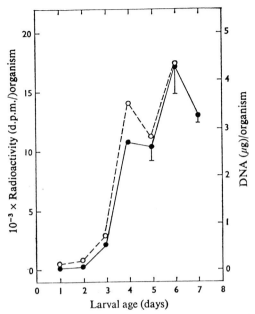

Fig. 3 *Two different measurements of DNA contents throughout the larval stage of the mosquito.*

The steady-state DNA contents (sDNA + pDNA) determined radiometrically by continuous labelling of larvae with [^{14}C] dT are plotted with the DNA contents determined by a colorimetric method (lang *et al.,* 1965). ●, Radiometric data; ○, colorimetric data.

Pulse-labelling - In general, rates of DNA synthesis in insects have been measured by radioautographic methods which cannot yield precise kinetic data. Therefore, we determined rates of DNA synthesis in mosquito larvae and adults by measuring the incorporation of a radioactive precursor after pulse-labelling. Initially, the kinetics of radiolabelling were investigated to determine the time-course and linearity of precursor incorporation in order to establish routine pulse times for subsequent experiments. In addition, it was necessary to select an appropriate concentration and specific activity of thymidine in the pulsing solution so that after processing there were high enough levels of radioactivity in the fractions for counting purposes.

Mosquito larvae must be pulsed by ingestion of the radioactive medium, but adult mosquitoes can also be injected. The routine pulsing procedures used to determine DNA synthetic rates are described as follows:

Rinsed larvae were pulsed routinely for 1 h by placing them into 10 μM-thymidine solution containing 1 μCi/ml of $[^{14}C]$thymidine. Adult female mosquitoes of known age were cold-inactivated before injection. One microliter of <u>Aedes</u> Ringer solution (0.13 M-NaCl/4.7 mM-KCl/1.9 mM-CaCl$_2$/ 7.0 mM-KH$_2$PO$_4$, adjusted to pH 6.8 with KOH), containing 0.002 μCi of $[^{14}C]$thymidine was injected into each mosquito through the membranous region between the first and second thoracic spiracles. Injection needles were made from Pasteur pipettes (Medglass) pulled out to very fine points (approx. 0.1 mm diam.) over a microflame. Injected mosquitoes were kept at 29° C for 45 ± 15 min and were then quickly inactivated at 0° C, weighed and homogenized. The precision of the procedure is indicated by the fact that the mean and S.E.M. of the injected radioactivity calculated from 26 separate experiments was 58717 ± 3614 d.p.m./ 100 mg body weight. The larval profiles of synthetic rates determined by pulse-labelling are shown in **Fig. 4**. The adult profiles of synthetic rates and also DNA contents are shown in **Fig. 5**.

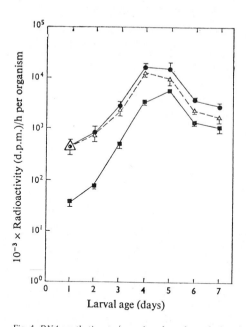

Fig. 4 *DNA synthetic rates/organism throughout the larval stage of the mosquito*

Axenically-cultured larvae were pulse-labelled 1h in [^{14}C] dT. The results, plotted semi-logarithmically, are expressed as radioactivity incorporated/organism. Each point with vertical bar represents the mean ± S.E.M. of eight to eleven samples from three or four different hatches. From 64 samples there was an overall recovery of radioactivity of 94.5 ± 0.96%. ●, sDNA + pDNA; △, pDNA; ■, sDNA.

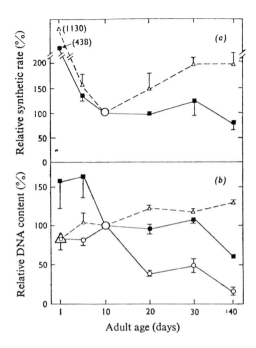

Fig. 5 *Changes in synthetic rates and contents of DNA throughout the adult stage of the mosquito.*
Synthetic rates were determined by pulse-labelling adult mosquitoes with [^{14}C]dT. Steady-state contents were determined from the incorporated radioactivity of mosquitoes which were continuously pre-labelled with [^{14}C]dT during larval development and allowed to develop into adults. The basis of expression was the percentage of homogenate radioactivity incorporated into acid-insoluble material. Relative rates and contents were based on the 10-day values of 100%. Each point with vertical bar represents the mean ± S.E.M. of three to nine samples. (*a*) Synthetic rates. The actual 10-day values/45 min were: pDNA, 1.47%; sDNA, 0.384%. Recovery of radioactivity was 99.3 ± 1.72% from 26 different samples. (*b*) Steady-state contents. The actual 10-day values were: pDNA, 67.7%: sDNA, 6.31%; acid-soluble fraction, 16.9%. Recovery of radioactivity was 92.1 ± 1.79% from 30 different samples. △, pDNA: ■, sDNA: ○, acid-soluble fraction.

Since the larvae ingested the medium, we found that about 45% of the homogenate radioactivity was from the gut contents. This gut–content radioactivity can effectively extend the pulse time and should be considered if a "chase" follows. Several changes of chase medium containing food particles will accelerate the elimination of the gut radioactivity. Although the chase medium should include non-radioactive thymidine, high concentrations should be avoided because the formation of excessive thymidine triphosphate is inhibitory to several enzymes which are involved in the biosynthesis of pyrimidine nucleotides and will thus inhibit the synthetic mechanism (38).

We have used the "pulse-chase" procedure to determine the metabolic fate of newly-synthesized DNA (34). Larvae were pulsed in $[^{14}C]$thymidine, transferred to media with 20,000-fold excess of non-radioactive thymidine, and sampled and processed at different time points during the next 24 hours to determine the acid-precipitable and acid-soluble radioactivities of the different fractions. The decrease in both the acid-soluble and sDNA fractions and concurrent increase in bulk DNA, pDNA, can be seen in Fig. 6.

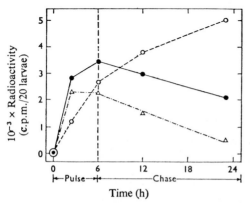

Fig. 6 *Incorporation of [^{14}C] thymidine into sDNA and pDNA in mosquito larvae*

Axenic larvae, 2 days old, were pulsed 6h in 20 ml of medium containing 20μCi of [^{14}C] thymidine (specific radioactivity, 45.5 mCi/mmol) and 'chased' with 20000-fold excess of non-radioactive thymidine. Larvae were sampled and processed at several time-points to determine the acid-precipitable and acid-soluble radioactivities of the different fractions. ●, sDNA; ○, pDNA; △, acid-soluble fraction.

Pulse-labelling techniques were also used to identify newly-synthesized DNA and to examine its subsequent increase in size during a chase. DNA from larvae sampled immediately after pulse-labelling and after one and two days of a chase was extracted by methods devised to provide maximal yield of DNA under conditions to inhibit endogenous, nuclease activity and was examined on linear sucrose gradients. After a 2 h pulse most of the labelled DNA had a sedimentation coefficient of about 8S in neutral gradients, but sedimented in a broad heterodisperse zone from 10-19S after a 4 h pulse (Fig. 7). Under alkaline conditions the peak fraction of labelled DNA had a sedimentation coefficient of around 8S after pulses up to 4 h duration, which indicated that the larger segments of DNA found at 4 h were not covalently linked (Fig. 8). Our results also

indicated that sDNA is not synthesized by the action of terminal deoxynucleo-
tidyltransferase which adds deoxyribonucleotides randomly to the ends of DNA
strands. This is shown in that after 0.5 - 6 h of pulse, single-stranded,
labelled DNA was found only in the 8S band and not in the 19-22S region of
larger DNA.

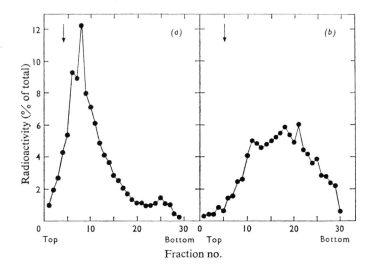

Fig. 7 *Neutral-sucrose-gradient centrifugation of pulse-labelled DNA*

Portions of DNA extracted from 3-day-old larvae which had been pulsed for 2 or 4h in 20 ml of medium
containing 20μCi of [^{14}C] thymidine (specific radioactivity, 51.8mCi/mmol) were analysed on neutral
sucrose gradients. Homogenates in a mixture of equal volumes of SSC medium and redistilled phenol
equilibrated with SSC medium were centrifuged at 19000g for 10min, and the upper aqueous phase was
removed and saved. The 'gummy' phenol phase was shaken again with a half volume of SSC medium and
re-centrifuged. The combined aqueous phases were extracted again with SSC medium/phenol, twice with
2 vol. of chloroform/butanol (24:1, v/v), and then incubated overnight at 37°C with 0.4mg of RNAase/
ml which had been pre-treated by heating at 80°C for 10min in 0.15M-NaCl, pH5.0. The solution was
applied to a column (30cm × 2.5cm) of Sepharose 6B and eluted with SSC medium. The first u.v.-ab-
sorbing peak, which contained the DNA, was concentrated by pervaporation, portions were sedimented
through neutral sucrose gradients, and the radioactivity of the fractions was determined. The position
of the haemoglobin marker (4.6S) is indicated by the arrows. (*a*) 2h pulse; total radioactivity 4075
c.p.m. (*b*) 4h pulse; total radioactivity 1491 c.p.m.

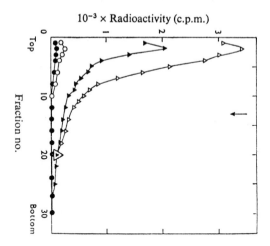

Fig. 8 *Alkaline-sucrose-gradient centrifugation of pulse-labelled DNA*

Axenic larvae, 3 days old, were pulsed as described in the legend to Fig. 5 for various times and then processed in alkali by a modification of the method of Okazaki *et al.*(1968). Larval homogenates in 0.3M-NaOH/0.01M-EDTA/6% (w/v) sodium *p*-aminosalicylate were incubated at 37°C for 1h and centrifuged at 3000*g* for 20 min. Portions of the supernatant fraction were sedimented through alkaline sucrose gradients and the acid-precipitable radioactivity of each fraction was determined. The position of marker, φX174 DNA (22.5S), is indicated by the arrow. ●, 0.5h; ○, 1h; ▲, 2h; △, 4h.

When larvae were pulsed for 6 h in $[^{14}C]$thymidine and "chased" with non-radioactive thymidine, the sedimentation coefficients of the labelled DNA in alkaline gradients increased to about 10-15S after 24 h and to 17-31S after 48 h (Fig. 9). Since the sedimentation coefficients of pDNA under alkaline conditions ranged from 25 to 30S, these experiments indicated that the labelled, newly-synthesized DNA consisted of small-size fragments which were later combined to form larger pieces in the size range of bulk DNA (pDNA).

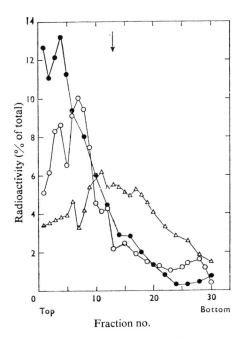

Fig. 9 *Alkaline-sucrose-gradient centrifugation of pulse-labelled and pulse-chased DNA*

Axenic larvae, 3 days old, were pulsed for 6h as described in the legend to Fig. 5 and then 'chased' for 1 or 2 days with a 1000-fold excess of thymidine. The processing conditions are described in the legend to Fig. 6. ●, 6h pulse, no chase; total radioactivity 9475 c.p.m.; ○, 6h pulse, 1 day chase; total radioactivity 6255 c.p.m.; △, 6h pulse, 2 day chase; total radioactivity 6220 c.p.m.

An additional tool to analyze the mechanism of DNA replication is the use of density-labelled analogs of thymidine such as bromodeoxyuridine (BrdUrd) to identify the newly-replicated DNA and facilitate its separation from "parental" DNA by equilibrium centrifugation. The in vivo labelling of mosquito larvae with sufficient bromodeoxyuridine proved to be a difficult problem until high concentrations were added to the pulsing medium. Although 25 mM caused a high degree of mortality, we could use 2.5 mM concentration level up to 24 hours without apparent harm. An interesting facet of these experiments was the demonstration that high concentrations of either bromodeoxyuridine or thymidine resulted in the preferential synthesis of the low molecular weight, sDNA and increased the sDNA/pDNA ratio more than 2 fold (Table 3).

| BrdUrd Concentration | pmol incorporated/4 h | | |
(mM)	sDNA	pDNA	sDNA/pDNA
25.0	24.6	18.4	1.34
10.0	17.9	8.51	2.10
1.00	1.50	1.00	1.50
0.10	0.092	0.137	0.672
0.01	0.014	0.024	0.583

Table 3. Effect of Bromodeoxyuridine Concentration on Its
Incorporation into DNA. 5 day larvae were pulsed for 4h in
different concentrations of BrdUrd containing [^{3}H] BrdUrd.
Samples were processed and counted by our standard procedure.

This procedure, in combination with double-labelling using two different
isotopes, is a powerful tool to discriminate between parental and de novo DNA's.
We have routinely cultured larvae in [^{14}C]thymidine media to label DNA continuously, "chased" out the radioactive nucleotide pool overnight and then pulsed
the larvae with either [^{3}H]thymidine or [^{3}H]bromodeoxyuridine. By this
double-labelling technique we found that sDNA is comprised of both parental and
de novo DNA. The exact nature of their relationship is still under investigation.

A word of caution must be noted concerning radiometric determinations.
It is most important in evaluating DNA metabolism to ascertain that observed
changes are inherent in the experimental organism and do not reflect changes in
an associated bacterial population. Mosquito larvae in nature, or grown on a
standard culture medium, derive their nutrients from bacterial infusions and
consequently they contain intestinal microflora. Bacteria did not present a
problem in experiments involving colorimetric determinations as their contri-
bution was negligible, but their presence did affect radiometric evaluations.
Comparisons were made between axenic- and standard-cultured (non-axenic)
mosquito larvae under different experimental conditions, and we found distinct
quantitative differences in uptake and incorporation of radioactive label by the
differently cultured larvae. These differences were present in continuously
labelled larvae from the same hatch as well as from several hatches (Fig. 10a).
In addition, pulsed axenic larvae exhibited a rate of incorporation three times

more rapid than the standard larvae and maintained linearity of both uptake and incorporation throughout a 4 h pulse (Fig. 10b). By contrast, standard larvae had a linear rate only up to 2 h. These findings formed the basis for using only axenically-cultured larvae in subsequent experiments. However, standard-cultured adult mosquitoes were satisfactory, since the precursors were administered by injection and thus circumvented the intestinal microflora.

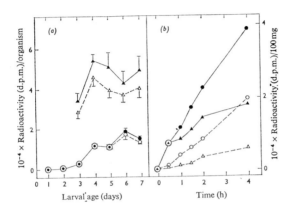

Fig. 10 *Differences in uptake and incorporation of [^{14}C] thymidine by continuously labelled or pulse-labelled axenic and standard (non-axenic)larvae*

Uptake is defined as the radioactivity in the homogenate, and incorporation is defined as the radioactivity in the acid-insoluble fraction. Samples were taken at the times indicated and processed as described in the Materials and Methods section. (a) Larvae from different hatches were continuously labelled by culturing under axenic or standard (non-axenic) conditions in media supplemented with [^{14}C]dT. Results are expressed as radioactivity (d.p.m.)/organism. (b) Larvae, 3 days of age, cultured under standard or axenic conditions were pulsed in $10 \mu M$ solutions of dT containing $1 \mu Ci/ml$ of [^{14}C] dT. Results are expressed as radioactivity (d.p.m.)/100mg larval weight. Each point with vertical bar represents the mean ± S.E.M. of five to nine samples for standard larvae, and of eight to eleven samples for axenic larvae. Bars are omitted if the S.E.M. is less than the size of the point. ▲, Standard uptake; △, standard incorporation; ●, axenic uptake; ○, axenic incorporation.

Earlier we noted that acid-insoluble fractions can be treated with DNAase to ascertain that the radioactive label was in DNA. It is important to confirm this fact, for if the label migrates due to metabolic conversion of the precursor, the radioactivity measurement will be misinterpreted. To identify the radioactivity with DNA, the samples are subjected to differential enzyme hydrolysis by successive digestions with Pronase, RNAase, and DNAase followed by acid precipitation at each step. If the radioactivity has been incorporated only into

DNA, the radioactivity will still be acid-insoluble after Pronase and RNAase
hydrolysis, but will become acid-soluble after DNAase treatment. An illustration
using Sephadex filtration instead of acid-precipitation is shown in Fig. 11.

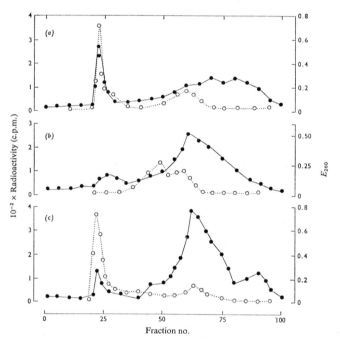

Fig.11 *Sephadex G-50 profiles of nuclease-treated supernatant fraction*

Portions of the supernatant fraction prepared from homogenates of [^{14}C] thymidine-labelled larvae were
applied to a column of Sephadex G-50 and the fractions were analysed for radioactivity and E_{260}. Samples;
(*a*) untreated; (*b*) DNAase-treated; (*c*) RNAase-treated. ○, Radioactivity; ●, E_{260}.

PURIFICATION OF DNA

The extraction and purification methods for DNA are almost as numerous
as its investigators, for each organism and tissue usually has specific limitations.
However, there are certain considerations important for experimental design as
well as several basic principles which underlie the different procedures. The
separation of nucleic acid components and specific methods of identification are
clearly outlined by Volkin and Cohn (39) and more recently Travaglini (5)
presented a survey of the general methodology for the extraction and purification
of DNA. Also, specific procedures for the isolation and analysis of DNA from

rat liver tissue and green algae have been outlined by Howell (19), and their
adaptation to insect tissues should be relatively simple. The reader is encouraged to refer to these excellent sources for more detailed information.

Three treatments form the basis of most DNA extraction procedures:
(a) lysis of cellular membranes to release DNA; (b) denaturation or inhibition
of nucleases; and (c) removal of proteins and other macromolecules. According
to Jordan (40) there is no absolute criterion for evaluating the purity of DNA
preparations, and the methods used, such as chemical analysis, molecular
weight and viscosity determinations, U.V. spectra, and acid-base titrations,
can only be of value for comparing different samples. Nonetheless, the absence
of protein, RNA and polysaccharides may be the most useful working criteria.

As Travaglini (5) points out, there are several popular methods that can
be used singly or in combination. The detergent method of Marmur (41) uses the
anionic detergent, sodium dodecyl sulphate (SDS), in an initial step to separate
proteins from the nucleic acids with additional separation and removal of proteins by the use of sodium perchlorate and Sevag treatment. The Sevag treatment (42) involves shaking the sample with an equal volume of chloroform-
isoamyl alcohol (24:1 v/v) and then centrifuging the mixture to separate the
aqueous and organic phases. After several such treatments, the DNA can then
be precipitated from the upper aqueous phase by the gentle layering of approximately 2 volumes of ethanol. When gently mixed with a stirring rod, the DNA
(and RNA) "spools" on the rod as a thread-like precipitate and is easily removed. The precipitate is redissolved in SSC, Sevag treated, and digested with
RNAase to remove the RNA. Additional Sevag treatments remove the RNAase
and then the DNA is selectively precipitated by isopropanol. In this method the
repeated shaking of DNA in solution subjects the long molecule to shear forces
and thus reduces the molecular weight. Using this procedure for bacterial cells,
Marmur reported DNA yields up to 50%.

The phenol method of Kirby (43, 44) depends upon the lipophilic and chelating properties of a salt anion (e.g., para-aminosalicylate) in combination
with phenol which apparently acts as a protein solvent. Isotonic salt solutions
or phenol alone do not release DNA, but Kirby found the combination always resulted in the extraction of DNA from mammalian cells. As an alternate to the
detergent method, phenol is probably more efficient in removing proteins and

requires fewer deproteinizing steps than chloroform-isoamyl treatment.

A common problem with either the phenol or the chloroform-isoamyl extraction method is the trapping of DNA in the interphase layer between the organic and aqueous phases. Beware of the possibility that newly-synthesized DNA may be preferentially localized in the interphase, according to different investigators (11,12,13,14).

To obtain pure DNA preparations it is necessary to remove RNA by RNAase digestion followed by Pronase digestion to remove the RNAase. RNA can also be hydrolyzed with strong alkali, but this procedure also denatures the DNA by disrupting the hydrogen-bonding between the two strands. Kirby reported an improved method using a phenol/m-cresol-8-hydroxyquinoline mixture which is a more powerful deproteinizing agent and also separates DNA from RNA without enzymatic treatment (45).

We have used adaptations of the Kirby phenol method to extract DNA in maximal yield for analysis on neutral sucrose gradients by adding phenol saturated with SSC to tissue homogenates or by homogenizing larvae in a mixture of equal volumes of phenol and SSC. On the other hand, to obtain pure preparations for physico-chemical studies, sDNA and pDNA were isolated from the supernatant and particulate fractions of larval homogenates by a modification of the Marmur method. This method, which is described below, yielded 0.17 mg of sDNA and 0.46 mg of pDNA/g of larvae. The concentration of total isolated DNA was, therefore, 0.63 mg/g of larvae, which was 63% of the value for total DNA in homogenates.

(a) sDNA - Sodium dodecyl sulphate was added to the soluble fraction to make a 2% (w/v) solution, was then incubated 10 min at 60° C and then shaken with an equal volume of chloroform/butanol (24:1, v/v) for 25 min. After centrifugation at 10000 g for 20 min the emulsion separated into an upper aqueous phase, a middle layer of denatured protein, and a lower chloroform phase. The upper phase was removed and shaken with chloroform/butanol at least six times until there was no visible middle layer of denatured protein. It was then incubated at 37° C for 1 h with 125 μg of heat-treated RNAase/ml, shaken four more times with chloroform/butanol, and extracted three times with an equal volume of ether. The solution was concentrated in dialysis tubing by pervaporation, applied to a column of Sepharose 6B and eluted with SSC medium. The first

UV -absorbing peak which contained the sDNA was concentrated, and the amount of sDNA was measured spectrophotometrically at 260 nm assuming an absorbance of 20 (in a 1 cm light path) for a 1 mg/ml solution. Ethanol treatment resulted in a flocculent precipitate of DNA.

(b) pDNA - The particulate fraction from the homogenate was resuspended in homogenizing medium, and sodium dodecyl sulphate was added to a concentration of 2% (w/v). The suspension was incubated at 60° C for 10 min. and then at 37° C for 2 h with 2 mg of preincubated Pronase/ml. It was then shaken at least seven times with chloroform/butanol, and the pDNA was then further purified as described for sDNA. Ethanol treatment resulted in fibrous DNA which could be "spooled."

DNA can be easily extracted from small quantities of tissue by equilibrium centrifugation of cell extracts in cesium salt solutions. Travaglini (5) reported a method by which Drosophila embryos were homogenized in 4 M CsCl and after centrifugation, the total nucleic acids and polysaccharides were recovered from a pellet. The pellet was then digested successively with alpha-amylase, RNAase and pronase.

We have purified radiolabelled DNA using cesium salt gradients by dissolving Cs_2SO_4 in cell extracts free of cellular debris and centrifuging the mixture to equilibrium in a fixed angle rotor. One of the important advantages of fixed angle rotors, as discussed by Flamm et al. (46), is the better separation of DNA from RNA since RNA pellets on the wall of the tube while DNA bands at its own density in the gradient.

In general, to band DNA, the initial density of the cesium salt solution should approximate the density of the insect DNA. For mosquito DNA we have used initial densities of $1.45-1.50 g/cm^3$ for Cs_2SO_4 and $1.70 g/cm^3$ for CsCl. In actual practice, 3.182 g of technical grade Cs_2SO_4 was added to 5.2 ml of a crude supernatant containing 1% Sarkosyl (final concentration) in a thick-walled, polyallomer tube which requires neither paraffin overlay or capping. For convenience, we have centrifuged for 60 h over a weekend. Each gradient was fractionated and portions were acid-precipitated and counted to locate the radioactive DNA. These peak fractions were pooled, dialyzed against SSC to remove the cesium salt and nucleotides, and concentrated by pervaporation. The A_{260}/A_{280} values for these DNA preparations were about 1.8 for pDNA and 1.6 for sDNA.

Although DNA bands in a cesium gradient of cell extract, a portion of DNA in the form of nucleoprotein may remain in a pellicle on top along with proteins and polysaccharides and should not be disregarded in recovery calculations.

The recovery of different functional macromolecules in addition to DNA may be an important feature of some studies. A non-enzymatic procedure to separate DNA, RNA and proteins from animal cells has been described recently by Shaw et al. (47). Molecular dissociation of nucleoprotein complexes is achieved by the use of a chaotropic salt, Na thiocyanate (NaSCN), and separation of DNA, RNA and protein is achieved by centrifugation in $CsCl - Cs_2SO_4$ density gradients.

Although hydroxyapatite (HAP) chromatography is commonly used to separate single- from double-stranded DNA (48,49,50), it has also been used to purify DNA from cell lysates. Britten et al. (51) demonstrated that in 8M urea and 0.014 M phosphate buffer, native DNA adsorbed to HAP while RNA passed through. The DNA was then recovered with a high degree of purity by elution with 0.4 M phosphate buffer. Tissues from all sources may not be equally suitable for this method, and DNA of high molecular weight may not adsorb completely to the column. Two recent modifications of the HAP method have been reported by Markov and Ivanov (52). Also Bernardi (53) has prepared a comprehensive discussion of hydroxyapatite chromatographic procedures.

CHARACTERIZATION OF DNA

The identification and characterization of a DNA species is dependent upon its physical and chemical properties. Most determinations of these characteristics require isolated, homogeneous DNA preparations that are free from RNA and protein. The key analysis is the base composition of DNA which can be determined in different ways. The molecular size and strandedness are additional important characteristics, especially in studies of DNA replication intermediates. Base Composition - A particular DNA can be characterized by the mole fractions of its four major purine and pyrimidine bases: adenine (A), cytosine (C), guanine (G), and thymine (T). This base composition can be expressed in several ways such as % of each base, % G + C, or A + T/G + C. These values are inter-

convertible since A = T and G = C due to base pairing. Thus, a given analytical
method can be validated by comparing the ratios, A + G/C + T, A/T, and G/C,
which should all have values of 1.0.

The base composition of DNA is diagnostic of the origin of the DNA and may
indicate phylogenetic relationships and evolutionary divergences. Differences
in composition exist not only between different phyla and genera as shown in
Table 4, but interspecific differences have also been demonstrated in Drosophila
(54). Our finding of the same base composition in both sDNA and pDNA provided evidence that sDNA was not a bacterial or viral contaminant which would
most likely have a different base composition (34).

DNA Source	Developmental Stage	% G + C	Reference
Insects:			
Apis mellifera	larvae, pupae	32	(74)
Hyalophora cecropia	adults	35	(74)
Aedes aegypti	larvae	39	(34)
Musca domestica	adults	39	(74)
Drosophila melanogaster	adults	40	(74)
Bombyx mori	larvae	40	(74)
Locusta migratoria	adults	41	(74)
Mammals:			
Human spleen		40	(64)
Calf thymus		42	(64)
Mouse spleen		45	(64)
Bacteria:			
Bacillus subtilis		35	(64)
Escherichia coli		50	(64)
Micrococcus lysodeikticus		72	(64)

Table 4. Base Composition of Insect and Other DNA's

The identification of different DNA species according to base composition
can also be applied at the subcellular level, and in some systems nuclear and
mitochondrial DNA's can be distinguished from each other (Table 5).

% G + C			
Source	Nuclear DNA	Mitochondrial DNA	References
Drosophila	39.7	25.5	(75, 55)
Ox	44.9	43.8	(19)
Chick	41.8	49.0	(19)
Yeast	40.8	25.5	(19)

Table 5. Base Composition of Nuclear and Mitochondrial DNA

Value for Drosophila nuclear DNA was from t.l.c. data. All other values
were calculated from buoyant density data.

Satellite DNA's are found in a number of organisms and may be separated
from bulk DNA in density gradients of cesium salts on the basis of their different
compositions. Nuclei of Drosophila embryos and also of Rhynchosciara adults
contain at least two satellite DNA's (54-57, 76). Also species differences in the
proportion of total DNA that is satellite DNA has been reported in Drosophila
(54) and in grasshopper (58). Satellite DNA has not been detected during our
studies of mosquito DNA's.

There are a variety of methods for the determination of the base composition
of DNA, and the most frequently used methods will be discussed including:
direct analysis of bases by thin-layer chromatography and indirect analyses by
buoyant density, spectral analysis and DNA melting temperature (T_m). It is
recommended that at least two methods be used to minimize the possibility of
error that may be inherent in any single method.

(a) Base analysis by thin-layer chromatography (t.l.c.) - Direct base
analysis is the standard of comparison for other methods. Both the original
paper chromatography method developed by Wyatt (59) and the t.l.c. adaptation
by Holdgate and Goodwin (60) worked well in our hands and an outline follows:

Purified DNA is acid-hydrolyzed to the free bases by hot perchloric acid.
The hydrolysate is pipetted onto cellulose thin-layer plates and developed with
an isopropanol-HCl-H_2O solvent system. Standard bases are run in parallel
lanes on the plate and ideally, the unknown samples should be bracketed by
standard samples. After development, the plates are dried, and the locations
of the standard and hydrolysate bases are revealed under UV light. Tentatively,

the bases are identified by comparison of their Rf values with standards.
Quantitation of the bases is achieved by scraping each spot from the plate,
eluting with acid and determining the molar concentration by the differential
extinction technique of Bendich (61). The identity of each base may be confirmed
by its unique UV spectrum at different pH's (62).

An alternative procedure is the enzymatic hydrolysis of DNA by DNAase to
nucleotides which are separated by t.l.c. on anion exchange plates (19, 63).
However, both methods should be validated by analysis of a DNA of known base
composition such as calf thymus or salmon sperm DNA which are commercially
available.

(b) Buoyant Density - DNA can be separated from other macromolecules in
cesium salt gradients on the basis of its specific buoyant density. This fact has
been used both for the purification of DNA as described in an earlier section,
and for the determination of its base composition.

When a solution of CsCl is centrifuged at high speeds, a concentration
gradient forms until an equilibrium is reached between centrifugal force and
the tendency of the CsCl to diffuse to a lower concentration. When DNA is
centrifuged in the CsCl solution it will band in the gradient where its buoyant
density equals the density of the CsCl solution.

It was demonstrated by Schildkraut et al. (64) that the buoyant density of
DNA in CsCl increases linearly with the % G + C content. Therefore, after
determining the buoyant density (ρ) of DNA in CsCl, the base composition can
be calculated according to the following equation:

$$\% \text{ G + C} = \frac{\rho - 1.660}{0.098}$$

This equation is valid for native, double-stranded DNA which contains only the
bases, A, C, G, and T, and has a G + C content between 20-80%. Furthermore,
the DNA preparation must be free of heavy metal ions.

The most precise analysis of density is obtained using isolated DNA in an
analytical centrifuge (46, 65), but satisfactory results can also be achieved in a
preparative centrifuge using either purified DNA or relatively crude preparations
of supernatants or subcellular fractions. The use of marker DNA's of known
densities is one of the easiest and most accurate methods to calibrate the
density gradient.

The initial density of the CsCl solution should approximate the density of the DNA to be analyzed. Most DNA's have densities ranging from 1.680 - 1.740 g/cm^3 in CsCl. Solid CsCl may be dissolved in a dilute DNA solution or a small volume of a concentrated DNA solution may be added to a CsCl solution of known density. To prepare the initial CsCl solution, either the weight per cent or the initial density must be known. A convenient equation is (66):

$$\text{Weight per cent} = 137.48 - 138.11 \ (\tfrac{1}{\rho})$$

The initial density of the CsCl solution is commonly determined from its refractive index using a refractometer, although the direct determination with a pycnometer can be used. Within the density range 1.25 - 1.90 g/cm^3, the following equation relating density (ρ) with refractive index (η) at 25° C can be used (66).

$$\rho = 10.8601 \eta \ - \ 13.4974$$

The initial density of the CsCl solution can be adjusted by adding either more CsCl or more diluting solution. The following equations can be used to calculate the required amounts of solid CsCl or diluent to change the density of a given volume (V) by a certain increment ($\Delta \rho$) (46).

$$\text{g (CsCl)} = \Delta \rho \ \text{x} \ \ V \ \ \text{x} \ 1.32$$
$$\text{ml (diluting solution)} = \Delta \rho \ \text{x} \ \ V \ \ \text{x} \ 1.52$$

When the CsCl is dissolved in a crude extract, or in the presence of high concentrations of urea, $NaClO_4$ or NaSCN, false density determinations may result from use of the refractive index.

Most cesium gradients are centrifuged 48-60 hours to reach equilibrium and usually, we have centrifuged the samples at 33,000 r.p.m. for about 60 hours at 20° C. At the end of the run, the rotor must decelerate without braking.

After centrifugation, the gradients are fractionated and analyzed for DNA by UV absorbance at 260 nm or by radiometric assay. At intervals throughout the gradient the densities of fractions should be determined in order to confirm the linearity of concentration. It is important that the DNA bands in the linear portion to determine its buoyant density.

Different aspects of the centrifugation of DNA in CsCl solutions have been comprehensively discussed by Flamm et al. (46). Also, cesium sulphate has been used instead of CsCl and the specific advantages of this salt have been out-lined by Szybalski (67).

(c) Spectral Analysis – The base composition of DNA can also be determined by UV analysis due to the unique spectra of its constituent bases. UV methods are rapid, simple and allow recovery of the sample. However, UV analysis usually requires isolated DNA that is free of other absorbing interferences such as RNA and protein.

The basic principle is that the absorbance ratios at different wavelengths (e.g., 245/270, 240/280, 240/275) are proportional to the % G + C. The methods of Hirschman and Felsenfeld (68) based on both native and denatured materials worked satisfactorily for mosquito DNA's. The newer procedure of Ulitzur (69) requires fewer measurements and has the advantages that the absorbance ratio 245/270 is essentially unaffected by the presence of protein. Further, this method is valid for DNA's with 31-72% G + C that are analyzed in 0.15-1.5 M salt concentrations at a temperature range of 5-35° C.

(d) Thermal Denaturation – As DNA is heated in solution, an increase in UV absorbance (hyperchromicity) occurs when the helical DNA is denatured to form a random coil (Fig. 12). The temperature at which this transition is 50% complete is termed the melting temperature (T_m). Marmur and Doty (70) showed that % G + C content over the range 25-75% is linearly related to the T_m and can be calculated by the following equation:

$$\% \text{ G} + \text{C} = \frac{T_m - 69.3}{0.41}$$

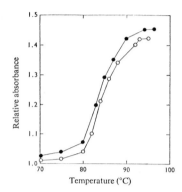

Fig. 12 *Thermal-denaturation profiles of purified mosquito DNA species*

The absorbance at each temperature, relative to the absorbance of the native material at 20°C, is plotted as a function of temperature. The T_m value of both DNA species was 84°C. •, sDNA; ○, pDNA.

For this procedure, DNA in SSC solution is heated slowly (1° C/15 min) to ensure thermal equilibration of the sample. Temperature can be regulated with a water bath connected to a jacketed cuvette holder, and the A_{260} at each temperature is corrected for the thermal expansion of water.

The gradual thermal transition curve of single-stranded DNA differs from that of native DNA and provides additional information for the characterization of a DNA species.

The base compositions of mosquito sDNA and pDNA were determined using the four methods described above, and the results are shown in Table 6. From 3-6 different DNA preparations were analyzed, and no difference was observed between the base composition of sDNA and pDNA regardless of method. Also the results from the different procedures agreed, for there was no statistical difference.

DNA Species	T_m	T. L. C.	% G + C Spectral Method	Buoyant Density
sDNA	36. 0	40. 0	37. 4	41. 3
pDNA	36. 0	40. 0	38. 4	38. 2

Table 6. Base Composition of Mosquito DNA's by Four Different Analyses.

UV Determinations of DNA - UV spectrophotometric analysis is one of the most useful procedures for DNA studies. In addition to base composition determinations, UV measurements are used to estimate the concentrations of aqueous DNA solutions, to determine the strandedness of DNA and to evaluate the purity of DNA in regard to protein. Since these are non-destructive procedures, they provide an additional advantage when there is a limited amount of material. One of the most common uses of UV measurements is to monitor absorbing materials during the elution of columns and fractionation of gradients.

The UV absorbancy of DNA is dependent upon the chromophoric property of its purine and pyrimidine bases, and a characteristic spectrum is shown in Fig. 13. Both the pH and ionic strength of the solution can affect the spectrum, and the sugar moiety plays a role if the pH is high enough to ionize the hydroxyl groups (71).

The absorbance of a neutral DNA solution at 260 nm can be used to approximate its concentration, assuming the DNA is double-stranded. Since histones absorb only slightly at 260 nm, the DNA content of chromatin solutions can also be estimated. A useful relationship is that the A_{260} of 1 mg/ml solution of DNA in a 1 cm light path is about 20. A necessary precaution in UV measurements of DNA is the absence of all other UV-absorbing substances such as nucleotides, coenzymes, RNA, proteins and certain chemical reagents including OH-ions and TCA. For this reason, PCA instead of TCA is often used to precipitate DNA.

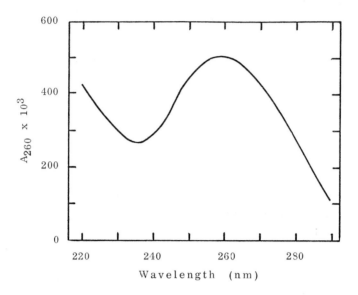

Fig. 13. Spectrum of mosquito pDNA

    If the DNA/RNA ratio exceeds 5, UV absorbancies at 260 and 280 nm can be used to evaluated the purity of DNA in regard to protein (72). Since proteins absorb maximally at 280 nm due to their tyrosine and tryptophan content, and the DNA maximum is at 260 nm, the A_{260}/A_{280} ratio will give an estimate of the percentage of nucleic acid in the sample, or conversely, indicate the amount of protein present. For example, with pure DNA, the A_{260}/A_{280} ratio is about 2.0 and the A_{280}/A_{260} ratio is about 0.5.

    Warburg and Christian (73) prepared various mixtures of isolated protein and RNA and measured their extinction coefficients at 280 and 260 nm, and from those values developed the following equation to determine the concentration of protein in the presence of nucleic acids:

$$\text{Protein concentration (mg/ml)} = 1.55 \, A_{280} - 0.74 \, A_{260}$$

Size Determination of DNA - Sucrose density centrifugation or velocity sedimentation is the most commonly used method to determine the sizes of DNA
molecules. In this procedure a solution of DNA is layered onto a preformed
sucrose gradient and centrifuged. The DNA sediments at a rate that depends
upon its size and thus larger compared to smaller particles move more rapidly
towards the bottom of the tube. The sucrose gradient technique provides a simple
means of estimating the size of an unknown DNA relative to that of a known DNA
by comparison of their sedimentation rates. The size of DNA is more often described by its sedimentation coefficient (S) than by molecular weight. One reason
is that the relationship between S and molecular weight is not exact since other
properties of DNA and the solvent affect sedimentation. Equations describing
these points are presented by Studier (77), and the basic principles of velocity
sedimentation are discussed in a number of articles (78-80).

Preformed, linear sucrose gradients are more generally used than "stepped"
gradients (81). Detailed procedures for the preparation and fractionation of preformed gradients have been described by McConkey (82), and a simplified method
to prepare several gradients simultaneously has been devised (83).

Before centrifuging samples it is wise to determine the linearity of some
test gradients. A simple way is to add nicotinamide to the sucrose solution in
one filling chamber, prepare and fractionate the gradient, and measure the absorbance at 260 nm of each fraction. A plot of absorbances versus fractions
should result in a straight line.

We have used 5-20% (w/v) gradients prepared from enzyme-grade sucrose
dissolved in 0.1 SSC/0.1 mM-EDTA for neutral gradients and in 0.1 N-NaOH/
0.9 M-NaCl/1.0 mM-EDTA for alkaline gradients. DNA samples (0.1 - 0.2 ml)
were carefully layered on top of 5.0 ml gradients and centrifuged in a swinging-
bucket rotor at 4° C. Although the time and speed of centrifugation are dependent upon the experiment, 15 hours at 25,000 r.p.m. was sufficient to move
purified sDNA and pDNA into the linear portion of the gradient.

Human hemoglobin, 4.6 S (84) and   φX174 DNA, 22.5 S, were used as
markers in the neutral and alkaline gradients, respectively. Under neutral conditions the DNA exists as double strands and in alkaline gradients, the DNA is
single-stranded. Fractions were collected from the top by displacing the gradient
with a 50% sucrose solution and were analyzed for UV absorbancy and radioactivity.

Sedimentation coefficients were calculated by using the equation of Martin and Ames (85).

When the size of DNA is to be determined. fresh tissues or homogenates should be used. A decrease in the size of frozen DNA compared to fresh DNA has been noted by us and others (5.6). although we do not know if the degradation is due to nuclease activity or shear by freeze-thaw.

A different method to determine sedimentation coefficients and molecular weights of DNA has been described by Taylor who used isokinetic gradients of $NaClO_4$ (86). Another method that is used for the separation and identification of different-sized DNA fragments is electrophoresis with polyacrylamide or agarose gels (87-89).

Determination of Strandedness - Since there is evidence that DNA replication is a discontinuous process which involves the synthesis of small pieces of DNA and their subsequent linkage to form bulk DNA, there has been much interest in the characterization of the intermediate fragments (78,50,90). One important aspect involves the strandedness of the DNA segments at different points in the replication process. The strandedness of DNA molecules has been investigated by hydroxyapatite chromatography (HAP), digestion by single-strand specific endonucleases. DNA melting curves and nitrocellulose adsorption. We have used HAP and the T_m procedures to study mosquito DNA, but suggest that other methods may be equally useful and perhaps more sensitive for the detection of single-stranded regions.

Since double-stranded DNA has a higher affinity for hydroxyapatite than single-stranded DNA, their separation can be achieved on the basis of their strandedness. Procedures for HAP chromatography have been well-covered by Bernardi (53), and our analysis which confirmed the double-stranded nature of sDNA utilized those methods. A batch method with HAP has recently been described for the quantitative determinations of small amounts of DNA (91). Before adopting the batch method. it is suggested that native and denatured DNA's from your experimental system be tested on a HAP column using a linear phosphate gradient in order to determine the critical molarities for their separate elutions. For example. under our conditions the DNA's eluted at lower molarities than those suggested by the authors.

Single-strand, specific endonucleases derived from Neurospora crassa and

Aspergillus oryzae are commercially available or can be prepared (92, 93).
These enzymes were found to be highly specific for the hydrolysis of single-
stranded DNA and have been used to detect the occurrence of single-stranded
fragments and also gaps in newly-synthesized DNA (89).

Single-stranded DNA preferentially adsorbs to nitrocellulose and this
characteristic is exploited to separate it from native DNA and to determine the
presence of single-stranded regions. The fractionation of nucleic acids can be
accomplished using either nitrocellulose membrane filters or columns (94, 95),
and both methods have been used for the analysis of newly-synthesized DNA
(14, 96). Membrane filter techniques have been particularly useful in experiments which exploit base-sequence homologies to form hybrid DNA-DNA or DNA-
RNA complexes (97. 98). Details of methods for hybridization or DNA reasso-
ciation kinetics are beyond the scope of this article. but the interested reader
may refer to other sources for this information (99-101).

CONCLUSIONS

This chapter was designed to provide basic information on methods needed
for initial studies of insect DNA in a wide range of investigations. In most instances, references have been included that should enable the reader to pursue
more detailed study, but no text can substitute for actual experience. It must be
pointed out again that the coverage has been restricted. and we have omitted
related areas such as DNA-protein interactions. transcription, and gene acti-
vation by hormones.

The study of insect DNA's is a relatively untapped area, and it was our
intention to encourage its exploration. but with proper regard for analytical
rigor and careful evaluation of methodology. These precautions are especially
important with insects because of the vast diversity of species and the complexity
of their developmental stages. On the other hand. these characteristics provide
distinct biological and technical advantages which make insects uniquely appropriate models for DNA studies.

REFERENCES

1   G. R. Wyatt, Biochem. J., 48 (1951) 584.

2   J. D. Watson and F. H. C. Crick, Nature, 171 (1953) 737.

3   C. A. Lang, H. Y. Lau and D. J. Jefferson, Biochem. J., 95 (1965) 372.

4   C. A. Lang and F. Meins, Jr., Proc. Nat. Acad. Sci. U. S., 55 (1966)
    1525.

5   E. C. Travaglini, in D. M. Prescott (Editor) Methods in Cell Biology,
    Vol. 7, Academic Press, New York, 1973, p. 105.

6   K. Yamafuji, F. Hashinaga and T. Fujii, Enzymologia, 31 (1966) 92.

7   G. A. Danielli and E. Rodino, Nature (London), 213 (1967) 424.

8   L. Tiepolo and A. Laudani, Chromosoma, 36 (1972) 305.

9   A. Krishnakumaran, H. Oberlander and H. A. Schneiderman, Nature
    (London), 205 (1965) 1131.

10 M. Locke, Tissue and Cell, 2 (1970) 197.

11 T. Ben-Porat, A. Stere and A. S. Kaplan, Biochim. Biophys. Acta, 61
    (1962) 150.

12 A. G. Levis, V. Krsmanovic, A. Miller-Faures, and M. Errera,
    Eur. J. Biochem., 3 (1967) 57.

13 D. L. Friedman and G. C. Mueller, Biochim. Biophys. Acta, 174 (1968)
    253.

14 H. Probst, A. Ullrich and G. Krause, Biochim. Biophys. Acta, 254 (1971)
    15.

15 R. L. Anderson and C. A. Lang, Gerontologist, 10 (1970) 24.

16 C. A. Lang, K. J. Basch and R. F. Storey, J. Nut., 102 (1972) 1057.

17 A. Zweidler and L. H. Cohen, J. Cell Biol., 51 (1971) 240.

18 W. C. Schneider and G. H. Hogeboom, J. Biol. Chem. 183 (1950) 123.

19 S. H. Howell, in M. Chrispeels (Editor) Molecular Techniques and
    and Approaches in Developmental Biology, John Wiley and Sons, New York,
    1973, p. 117.

20 J. Wilcockson, Anal. Biochem. 66 (1975) 64.

21 H. N. Munro, in D. Glick (Editor) Methods of Biochemical Analysis,
    Vol. XIV, Interscience, New York, 1966, p. 113.

22 K. Burton, Biochem. J., 62 (1956) 315.

23 R. B. Church and F. W. Robertson, J. Exp. Zool., 162 (1966) 337.

24 B. Linzen and G. R. Wyatt, Biochim. Biophys. Acta, 87 (1964) 188.

25 A. Devi, A. Lemonde , U. Srivastava and N. K. Sarkar, Exptl. Cell Res.,
    29 (1963) 443.

26 G. M. Price, J. Insect Physiol.. 11 (1965) 839.

27 G. M. Richards, Anal. Biochem., 57 (1974) 369.

28 J. M. Webb and H. B. Levy, J. Biol. Chem., 213 (1955) 107.

29 R. F. Martin, D. C. Donohue and L. R. Finch, Anal. Biochem., 47 (1972)
    562.

30 S. Lovtrup, Acta Biochimica Polonica, 9 (1962) 411.

31 J. M. Kissane and E. Robins, J. Biol. Chem. 233 (1958) 184.

32 D. Roberts and M. Friedkin, J. Biol. Chem., 233 (1958) 483.

33 P. C. Beers and J. L. Wittliff, J. Biol. Chem., 233 (1975) 483.

34 P. C. Kao, C. F. Beyer and C. A. Lang, Biochem. J., 154 (1976) 471.

35 B. J. Mills and C. A. Lang, Biochem. J., 154 (1976) 481.

36 K. Selman and F. C. Kafatos, J. Insect Physiol., 20 (1974) 513.

37 K. D. Chaudhary and A. Lemonde, Canad. J. Biochem., 44 (1966) 1571.

38 J. E. Cleaver (Editor) Frontiers of Biology, Vol. 7, North-Holland,
    Amsterdam, 1967, p. 51.

39 E. Volkin and W. E. Cohn in D. Glick (Editor) Methods of Biochemical
    Analyses, Vol. 1, Interscience, New York, 1954, p. 287.

40 D. O. Jordan, The Chemistry of Nucleic Acids, Butterworths, Washington,
    1960, p. 8.

41 J. Marmur, J. Mol. Biol. 3 (1961) 208.

42 M. G. Sevag, D. B. Lackman and J. Smolens, J. Biol. Chem., 124 (1938)
    425.

43 K. S. Kirby, Biochem. J., 66 (1957) 495.

44 K. S. Kirby, Progress in Nucleic Acid Research, 3 (1964) p. 1

45 K. S. Kirby, Biochem. J., 104 (1967) 254.

46 W. G. Flamm, M. L. Birnstiel and P. M. B. Walker, in G. D. Birnie
    (Editor) Subcellular Components, University Park Press, Baltimore, 2nd
    Ed., 1972, p. 279.

47 J. L. Shaw, J. Blanco and G. C. Mueller, Anal. Biochem., 65 (1975) 125.

48 J. F. Habener, B. S. Bynum and J. Shack, J. Mol. Biol., 49 (1970) 157.

49 G. J. Hayton, C. K. Pearson, J. R. Scaife and H. M. Keir, Biochem. J., 131 (1973) 499.

50 H. V. Hershey and J. H. Taylor, Exp. Cell Res., 85 (1974) 79.

51 R. J. Britten, M. Pavich and J. Smith, Carnegie Inst. Washington Yearb., 68 (1970) 400.

52 G. G. Markov and I. G. Ivanov, Anal. Biochem., 59 (1974) 555.

53 G. Bernardi, in L. Grossman and K. Moldave (Editors) Methods in Enzymology, Vol. 21, Part D, Academic Press, New York, 1971, p. 95.

54 C. D. Laird and B. J. McCarthy, Genetics, 60 (1968) 303.

55 E. C. Travaglini and J. Schultz, Genetics, 72 (1972) 441.

56 M. Blumenfeld and H. S. Forrest, Proc. Nat. Acad. Sci. USA, 68 (1971) 3145.

57 C. S. Lee, Biochemical Genetics, 12 (1974) 475.

58 I. Gibson and G. Hewitt, Nature, 225 (1970) 67.

59 G. R. Wyatt, Biochem. J., 48 (1951) 584.

60 D. P. Holdgate and T. W. Goodwin, Biochim. Biophys. Acta, 91 (1964) 328.

61 A. Bendich, Methods Enzymol., 3 (1957) 715 (Acad. Press, New York).

62 D. B. Dunn and R. H. Hall, in H. A. Sober (Editor) Handbook of Biochemistry, The Chemical Rubber Company, Cleveland, Ohio, 1968, p. G3.

63 K. Randerath and E. Randerath, Methods Enzymol., 12A (1967) 323 (loc. cit.)

64 C. L. Schildkraut, J. Marmur and P. Doty, J. Mol. Biol., 4 (1962) 430.

65 M. Mandel, C. L. Schildkraut and J. Marmur, Methods Enzymol., XIIB (1968) 184, Acad. Press, New York.

66 J. Vinograd and J. E. Hearst, Fortschritte der Chemie Organischer Naturstoffe, 20 (1962) 373.

67 W. Szybalski, in L. Grossman and K. Moldave (Editors) Methods in Enzymology, Vol. 12, Part B, Academic Press, New York, 1968, p. 330.

68 S. Z. Hirschman and G. Felsenfeld, J. Mol. Biol., 16 (1966) 347.

69 S. Ulitzur, Biochim. Biophys. Acta, 272 (1972) 1.

70 J. Marmur and P. Doty, J. Mol. Biol., 5 (1962) 109.

71 W. E. Cohn, in S. P. Colowick and N. O. Kaplan (Editors) Methods in Enzymology, Vol. 3, Academic Press, New York, 1957, p. 738.

72   J. Bonner, G. R. Chalkley, M. Dahmus, D. Fambough, F. Fujimura,
     R. C. Huang, J. Huberman, R. Jensen, K. Marushige, H. Ohlenbusch,
     B. Olivera and J. Widholm, in L. Grossman and K. Moldave (Editors)
     Methods in Enzymology, Vol. 12, Part B, Academic Press, New York,
     1968, p. 3.

73   O. Warburg and W. Christian. Biochem. Z., 310 (1941) 384.

74   S. H. Shapiro, in H. A. Sober (Editor) Handbook of Biochemistry, The
     Chemical Rubber Company, Cleveland, Ohio, 1968, p. H-43.

75   M. P. Argyrakis and M. J. Bessman, Biochim. Biophys. Acta, 72 (1963)
     122.

76   R. A. Eckhardt and J. G. Gall, Chromosoma, 32 (1971) 407.

77   F. W. Studier. J. Mol. Biol., 11 (1965) 373.

78   E. Burgi and A. D. Hershey. J. Mol. Biol., 3 (1961) 458.

79   E. Burgi and A. D. Hershey, Biophys. J., 3 (1963) 309.

80   J. Abelson and C. A. Thomas. Jr., J. Mol. Biol., 18 (1966) 262.

81   R. J. Britten and R. B. Roberts. Science, 131 (1960) 32.

82   E. H. McConkey, in L. Grossman and K. Moldave (Editors) Methods in
     Enzymology, Vol. 12, Part A, Academic Press, New York, 1967, p. 620.

83   A. B. Stone, Biochem. J., 137 (1974) 117.

84   T. M. Jovin, P. T. Englund and A. Kornberg, J. Biol. Chem., 244 (1969)
     3009.

85   R. G. Martin and B. N. Ames. J. Biol. Chem., 236 (1961) 1372.

86   J. H. Taylor. A. G. Adams and M. P. Kurek. Chromosoma, 41 (1973) 361.

87   S. Gregson. Anal. Biochem., 48 (1972) 613.

88   P. G. N. Jeppesen. Anal. Biochem., 58 (1974) 195.

89   J. R. Gautschi and J. M. Clarkson. Eur. J. Biochem., 50 (1975) 403.

90   M. Oishi. in L. Grossman and K. Moldave (Editors) Methods in Enzymology,
     Vol. 21. Part D. Academic Press. New York, 1971, p. 304.

91   V. Paetkau and L. Langman, Anal. Biochem., 65 (1975) 525.

92   S. Linn and I. R. Lehman. J. Biol. Chem., 240 (1965) 1287, 1294.

93   V. M. Vogt. Eur. J. Biochem. 33 (1973) 192.

94   A. P. Nygaard and B. D. Hall. Biochem. Biophys. Res. Commun., 12
     (1963) 98.

95 J. A. Boezi and R. L. Armstrong in L. Grossman and K. Moldave (Editors)
    Methods in Enzymology, Vol. 12, Part A, Academic Press, New York, 1967,
    p. 684.

96 H. Berger, Jr., and J. L. Irvin, Proc. Nat. Acad. Sci. U.S., 65 (1970) 152.

97 D. Gillespie in L. Grossman and K. Moldave (Editors) Methods in Enzy-
    mology, Vol. 12, Part B, Academic Press, New York, 1968, p. 641.

98 M. L. Pardue and J. G. Gall in D. M. Prescott (Editor) Methods in Cell
    Biology, Vol. 10, Academic Press, New York, 1975, p. 1.

99 R. B. Church in M. J. Chrispeels (Editor) Molecular Techniques and
    Approaches in Developmental Biology, John Wiley and Sons, New York,
    1973, p. 223.

100 D. W. Smith in M. J. Chrispeels (Editor) Molecular Techniques and
    Approaches in Developmental Biology, John Wiley and Sons, New York,
    1973, p. 165.

101 R. J. Britten, D. E. Graham and B. R. Neufeld, in L. Grossman and
    K. Moldave (Editors) Methods of Enzymology, Academic Press, New York,
    Vol. 29, 1974, p. 363.

ACKNOWLEDGMENTS

We are indebted to The Biochemical Journal for their kind permission to
reproduce Figs. 2-12 and other material which originally appeared in that
journal. Also we are grateful to the U. S. Public Health Service, National
Science Foundation, and Sigma Kappa Foundation for research grant support
and for a NIH Research Career Development Award to C.A.L.

CHAPTER 3

PREPARATION AND ANALYSIS OF RNA

BRADLEY N. WHITE

Department of Biology, Queen's University, Kingston, Ontario, Canada.

FERNANDO L. DE LUCCA

Departmento de Bioquimica, Faculdade de Medicina de Ribeirao   Preto da

Universidade de Sao Paulo, Ribeirao Preto, S.P. Brazil.

CONTENTS

I   INTRODUCTION

This chapter is designed for those who are not experienced with RNA but who find they need to estimate or isolate and analyse RNA in order to answer particular questions in their systems. We have tried to point out the special problems that may be encountered with insect tissues and the types of modification that need to be made to procedures that have been established with vertebrate material. This chapter makes no attempt to review all the techniques used for RNA analysis in insects but concentrates on some approaches that we have found to be useful.

While molecular approaches to the vital questions in insect biology lag behind the advances through the use of these techniques in vertebrate systems, rapid progress is being made in this direction. Indeed exploitation of several insect systems has had significant impact on molecular biology in general. The analysis of the tRNAs and isolation and sequencing of fibroin mRNA from the silkgland of Bombyx mori is one such system [1-5]. Others include the cocoonase and chorion protein systems of Antherea polyphemus and Manduca sexta [6-8]. It has been recognised for some time that <u>Drosophila</u> <u>melanogaster</u> with its battery of mutants and its polytene chromosomes might provide an excellent organism to study such questions as the structure of the eukaryotic genome and control of genes during development. Recent new elegant approaches have confirmed this [9].

II GENERAL COMMENTS

The basic problems in the isolation and analysis of RNAs from insects are clearly the same as those encountered with other organisms. These are the removal of the RNA from the cell and separation from other molecules while protecting the RNA from possible denaturing and degradative conditions. Although the properties of the various types of RNA found in insects are not very different from those from other sources, differences do occur in the nature and amounts of contaminating substances, as well as in the amounts of degradative enzymes. The extent of the analysis desired will clearly influence the procedures adopted to overcome the special difficulties encountered in insect cells. For example, in the estimation of the content of total RNA a major problem might be uric acid contamination, while during the isolation of biologically active messenger RNA the major problem to overcome will be ribonuclease activity.

Most cells contain ribosomal RNA (rRNA), transfer RNA (tRNA) and messenger RNA (mRNA) in roughly a 100:10:1 ratio. This also approximates the relative ease of extraction of these molecules as well as the approximate relative yields. Clearly the isolation of rRNA will require less effort than the isolation of mRNA.

Insect ribosomes contain rRNA components designated 26S, 18S and 5S [10-13]. The rRNA from the larger ribosomal subunit is slightly smaller than that found in vertebrates. Insect rRNA varies widely in its base composition and Diptera have rRNA distinguished by a very low G + C content [14].

The mRNAs are found mainly in the polysomes and comprise no more than 3% of the total RNA. Their size distribution is heterogeneous. The most important property of this type of RNA, in terms of its isolation, is the presence of a polyadenylate (poly A) segment (usually 150-200 nucleotides) which has been found in most of the eukaryotic mRNAs so far examined, except those coding for histones [14, 15]. It has been found recently that insect mRNAs have relatively short poly(A) sequences [3,18]. Although the precise function of the poly(A) fragment remains to be established, it is known that deadenylated mRNA can be faithfully translated in a cell-free protein synthesising system [19, 20].

The tRNAs are about 70-80 nucleotides long, corresponding to a molecular weight of 25,000 and a sedimentation coefficient of 4S. Each tRNA molecule is specific for a particular amino acid. The aminoacyl-tRNA synthetases (at least one for each amino acid) specifically esterify amino acids to the tRNAs and this serves as the best biological assay for these molecules. Because of their similar size and structural relationships, the different tRNA species are not easily separated. However, newer ion exchange chromatographic procedures have now made tRNA analysis routine. One of the more interesting features of tRNAs are the numerous modified bases found in these molecules. Analyses of these bases in purified tRNAs have revealed interesting correlations with the nature of the codons the tRNAs recognise.

III   ESTIMATION OF TOTAL RNA

It is often necessary to determine the changes in total RNA under various conditions or at different developmental stages [21, 23]. This is best achieved by a modification of the method of Fleck and Munro [24], which depends on direct measurement of RNA by its absorbance of UV light at 260 nm. Nucleotides and other UV absorbing molecules of low molecular weight are first removed by extraction with cold $HClO_4$, then the RNA is hydrolysed by alkali which renders it acid-soluble, while DNA resists the hydrolysis and remains acid insoluble. The only real difficulty encountered with this technique is the presence in some insect tissues of high levels of uric acid which absorbs UV light and has only low solubility in $HClO_4$ [25, 26]. RNA content can also be determined by the orcinol procedure [27-30] or by phosphate determination [31, 32] but these techniques also suffer from interference by contaminating molecules and are more time consuming than the direct measurement of absorbance at 260 nm [24].

Procedure

The tissue is homogenised in 20 volumes (v/w) of ice cold 0.5 $\underline{N}$ $HClO_4$ for 1 min. This is allowed to stand 10 min at $0^\circ C$, centrifuged and the supernatant discarded. The precipitate is washed at least four times with 0.2 $\underline{N}$ $HClO_4$. The supernatant should be monitored and washing continued until it contains negligible A_{260} material [25]. This will indicate that acid soluble substances such as uric acid which would interfere with the spectrophotometric determination have been removed.

The precipitate is further washed with 95% ethanol once and ethanol:ether
(3:1) twice and finally with ether twice. The precipitate is then dissolved
in 0.3 $\underline{N}$ KOH, incubated at 37°C for 1 h, cooled in ice and the protein and
DNA precipitated by adding 0.6 volumes 1.2 $\underline{N}$ $HClO_4$. The alkali-digested
RNA products are now acid-soluble. Incomplete hydrolysis can sometimes be
a problem, which can be overcome by longer hydrolysis times or a higher
ratio of 0.3 $\underline{N}$ KOH to precipitate. After standing for 10 min at 0°C, during
which time the $HClO_4$ will separate in crystalline form, the material is
centrifuged and the supernatant decanted and retained. The precipitate is
washed with 0.2 $\underline{N}$ $HClO_4$, centrifuged and the 2 supernatants combined. This
solution containing the ribonucleotides is diluted with water to make 0.1
$\underline{N}$ $HClO_4$. It may be further diluted with 0.1 $\underline{N}$ $HClO_4$ as necessary for
measurement of A_{260}.

The UV absorption is read at 260 nm in a quartz cuvette of 1 cm path
length. An absorbance of 1.0 at 260 nm is approximately equivalent to 32
µg RNA/ml. An UV spectrum of this material will confirm the purity of the
RNA. Uric acid is indicated by a shoulder on the spectrum in this region.

IV   USE OF ISOTOPES TO MEASURE RNA SYNTHESIS AND TO LABEL RNA

When only small amounts of tissue are involved or when determination
of rates of RNA synthesis are required, the analysis is best achieved with
labeled RNA precursors. Label can be introduced by injection, feeding or
addition to the medium when tissue or organ culture is being used. The
labeled precursor can be specific to RNA, such as [14]C- or ^{3}H-labeled
uridine, or more general such as orotic acid (an intermediate in pyrimidine
biosynthesis) or ^{32}P-labeled phosphate. After labeling, the tissue can be
treated in the same way as described above for unlabeled material to ensure
that the isotope measured is being incorporated into RNA rather than
another macromolecule. It must be noted that nucleotides do not easily
cross cell membranes so that RNA precursors are introduced in the form of
the bases or more usually as nucleosides. Absolute rates of synthesis can
only be obtained when the kinetics of incorporation are followed using
different concentrations of the labeled precursor and an estimate of
internal pool size made. A convenient method for estimating rates of
synthesis, specific activity and internal pool sizes is the one described
by Emerson and Humphreys [33]. This general approach was successful in
determining the rate of RNA synthesis in Drosophila embryos [34]. Similar
determinations have been made in organ culture [35] and in cell cultures

[36, 37].

Labeling becomes very important in the analysis of mRNA as this represents such a small fraction of cellular RNA. Injection of ^{32}P-labeled inorganic phosphate and labeling during organ culture were used to isolate and analyse the moth chorion protein mRNAs [38]. Labeling with adenosine and uridine have been used to study the processing of mRNAs in Drosophila and Aedes cell culture [38]. Labeling in vivo has been used to prepare RNA for RNA-DNA hybridisation procedures [39]. Use of isotopes can therefore be coupled to the isolation techniques described below to allow experiments to be performed on small amounts of tissue or to produce labeled material for various purposes.

V ISOLATION OF RNA - GENERAL

This section deals with procedures used during the extraction of the RNA from cells or tissues. The actual approach employed in any given situation will be dictated by the type of RNA that is being isolated as well as certain properties of the tissue itself, such as ribonuclease content. The general procedure involves the disruption of the cells, dissociation of the RNA from proteins and the precipitation and removal of the proteins from the nucleic acid. The first decision to be made is the nature of the solution in which the tissue is to be homogenised. During this process the RNA will be first exposed to ribonucleases and other possible denaturing conditions. Properties of the extraction buffer, presence of ribonuclease inhibitors and of dissociating and deproteinising agents are therefore important.

Properties of the extraction buffer

The buffer should be in the range of pH 5-9. We have found that pH 5 is effective in reducing ribonuclease activity in adult Drosophila but this may not necessarily be the case in other insects. The pH of the buffer can be very important in the isolation of mRNA because it can determine whether the poly(A) tail will bind to denatured proteins during deproteinisation; this will be discussed further under the section on mRNA.

The presence of some monovalent cations is important to maintain secondary structure of RNA and reduce its susceptibility to ribonuclease. Divalent cations are also important in maintaining secondary structure and reducing ribonuclease susceptibility, especially for the tRNAs. However, divalent cations can also lead to aggregation which will result in lowered yields.

Ribonuclease Inhibitors

        The source of ribonucleases during the extraction can be either
from the tissue itself (endogeneous) or from the glassware and solutions
(exogenous). The introduction of exogenous ribonucleases can be minimized
by a few simple precautions. Glassware and solutions should be autoclaved.
The glassware can be treated for 15 min with a 0.1% solution of diethyl
pyrocarbonate (DEP-see below). The treated glassware is then dried at
150°C for three hours to ensure complete breakdown of the DEP. Similarly,
solutions can be treated with 0.05% DEP for a few minutes at room
temperature and then boiled for 15-30 minutes to ensure complete breakdown
of the DEP. However, treatment with DEP must be avoided for certain
buffers, such as Tris, with amino groups which will react with DEP.

        Ribonuclease inhibitors fall roughly into three categories; those
which bind to ribonucleases and render them insoluble; those which react
with and inactivate ribonucleases; and those which inhibit the activity of
ribonucleases. The choice for addition to the extraction buffer will
depend on the RNA being isolated and the purpose for which it is to be
used.

1) Bentonite.   This has been used extensively to reduce the action of
ribonucleases in preparations from eukaryotes including insects [12,40,41].
It is a montmorillonite clay, $Al_2O_3.4 SiO_2.H_2O$, possessing sites which bind
to basic proteins such as many ribonucleases [42, 43]. The powder is
usually washed first with a solution of EDTA to remove metal ions and the
particle size selected by differential centrifugation. The suspension is
added to the extraction medium and finally removed together with the bound
ribonucleases by centrifugation. Bentonite also adsorbs RNA and we have
found it can markedly reduce yields if used in excess.

2) Diethyl pyrocarbonate (DEP). $C_2H_5O-CO-O-CO-O-C_2H_5$ is highly unstable
in the presence of water and reacts rapidly with lysine ε-amino groups and
tryptophan residues in proteins [44]. It also reacts with RNA, especially
adenine bases [45] which can lead to loss of biological activity [46,47].
Its major use is in the inactivation of nucleases on glassware and in
solutions rather than nucleases from the tissue itself.

3) Heparin.   This sulfated polysaccharide has been successfully employed
in reducing ribonuclease activity in many eukaryotes including insects
although its mode of action is not understood. We have used heparin in the
homogenising buffer (500 μg/ml) and in sucrose solutions (80 μg/ml) during
the preparation and analysis of polysomes from locust fat body but have
found that it is not effective in Drosophila. Heparin is a potent

inhibitor of in vitro protein synthesising systems [48] and must be removed
if active mRNA preparations are required. This is usually achieved by
sucrose gradient centrifugation.

4) Polyvinyl sulfate.   This is also a polyanion and appears to be able
to inhibit ribonucleases because it can bind basic proteins. It is active
at low concentrations and was effective at 2 µg/ml in the extraction of
biologically active mRNA from the wings of the silkmoth Hyalophora
cecropia [49]. We routinely add this to extraction buffers for Drosophila
RNA. The main drawback to this inhibitor is the difficulty of removal
from RNA preparations.

5) Rat Liver Supernatant.   Rat liver contains a protein which acts as a
natural ribonuclease inhibitor [50, 51]. The supernatant obtained by
centrifuging the post-mitochondrial fraction at 105,000 g for 4 h can be
used as a crude source of the inhibitor [52]. It is not yet established
whether this is effective with insect ribonucleases. It appeared not to
be helpful in preserving intact polysomes from the epidermis of H. cecropia
but that this tissue may contain an inhibitor of its own [53].

6) Others.   The action of ribonucleases during RNA isolation can be
reduced by using antisera against the enzymes [54], but this usually
entails first preparing purified ribonucleases. The addition of proteases
has also been used to minimize ribonuclease activity [55]. A copolymer of
L-tyrosine and L-glutamic acid has also been used [56]. Macaloid, a
purified hectorite (sodium magnesium lithofluoro silicate) has been used
in much the same way as bentonite [57, 58]. Various agents which effect
the deproteinisation of RNA also serve as ribonuclease inhibitors and
these will be discussed later.

Dissociation and deproteinisation agents

        These agents can be added to the extraction buffer before
homogenisation of the tissue and may help in the lysis of the cells or
they can be added after cell disruption. The presence of these agents
during the homogenisation can reduce the activity of ribonucleases and
their addition is therefore essential in tissues with a high activity of
these enzymes.

Dissociating agents - anionic detergents

        The presence of these dissociating agents helps in both cell and
organelle lysis as well as removal of the protein from the nucleoprotein
complexes in which the rRNAs and mRNAs are found. These detergents

include sodium dodecyl sulfate [59, 60] naphthalene 1,5-disulfonate
[61, 62], sodium 4-aminosalicylate [63], and sodium tri-isopropylnaphtha-
lene sulfate [63]. Sodium dodecyl sulfate (SDS, $CH_3(CH_2)_{11}SO_4Na$), the
most widely used detergent combines a non-polar hydrophobic group with
a strongly hydrophilic sulfate group. It is suitable, therefore, for
complex formation both with non-polar side chains and charged groups of
amino acids in proteins. The binding of SDS to proteins causes
denaturation. The detergent can be added to the extraction buffer
initially, or after homogenising the sample, to give a final concentration
of 0.5-2%. In preparing RNA from whole cells, when contamination with DNA
can be a serious problem, it is desirable to keep the SDS concentration
down to 0.5% and thus maintain the DNA-protein complexes while still
releasing the RNA.

  The main disadvantages of SDS are that it precipitates in the cold
and in the presence of potassium ions. The presence of potassium in the
extraction medium should be avoided. Sucrose gradients containing SDS
must be centrifuged at 16-20°C. Lithium dodecyl sulfate (LDS) is equally
efficient as SDS and does not precipitate at low temperatures [60].
Sodium dodecyl sarcosinate (Sarkosyl) is also more soluble than SDS at low
temperatures, but forms insoluble magnesium and manganese salts and
moreover it absorbs UV light at 260 nm [64].

Deproteinisation agents
  The anionic detergents used to dissociate the nucleoprotein complex
cause denaturation of the proteins but leave them soluble. Removal of
the proteins from the RNA is achieved by deproteinising agents which
precipitate them and allow separation by centrifugation, to give a
gelatinous precipitate at the interphase.
1. Phenol. The pioneering studies of Gierer and Schramm [65] and Kirby
[61] in 1956 showed that phenol denatures and extracts the cellular
proteins while nucleic acids can be recovered in their native form. After
20 years phenol is still the most widely used deproteinising agent. Water-
saturated phenol is a colorless liquid at room temperature but is solid at
lower temperatures. It becomes oxidized in contact with air and this can
be easily detected because of the yellow color of these products. To
prevent this, some batches of phenol are commercially supplied with
preservatives. It is advisable to redistill the phenol and store it in
brown bottles in the refrigerator.
  Addition of phenol to the extraction buffer forms a two-phase

system. Usually an equal volume of buffer-saturated phenol is added before
or after homogenisation of the tissue and the emulsion shaken to ensure
denaturation of the proteins. The separation of the phases is then usually
achieved by centrifugation at about 8,000 g for 10 min, which yields an
upper aqueous phase containing the RNA and variable amounts of contaminants
such as DNA, glycogen and other polysaccharides. The denatured proteins
are found in the lower phenol phase as well as in the gel present in the
interphase. Some RNA also remains bound to the protein gel at the
interphase [16, 66, 67]. Special emphasis will be given to this problem
in the section on mRNA, and other specific problems such as DNA and
carbohydrate contamination will be discussed in the tRNA and rRNA sections.

The efficiency of phenol deproteinisation can be improved by using
certain additives:

a) 8-hydroxyquinoline. This is usually used at a concentration of 0.1%
in order to prevent the oxidation of the phenol and reduce RNA-protein
interactions through divalent cations, which it chelates. It also serves
to inhibit the activity of ribonucleases [68].

b) m-cresol. This can be added to the phenol to a final concentration
of 10% both to lower the freezing point of the phenol and to improve the
efficiency of deproteinisation [63].

c) chloroform. This is usually added to give a 1:1 (v/v) phenol:
chloroform mixture and its presence increases the density of the phenol
phase which results in a better separation of the phases. Its use in mRNA
extraction is very important and will be discussed later.

2. Chloroform. Chloroform can be used in the absence of phenol as a
deproteinising agent in a procedure originally developed for the isolation
of DNA [69] and is especially efficient as a mixture with isoamyl alcohol,
added as an antifoaming agent [70-72]. Chloroform is particularly useful
in the extraction of tissues with a high content of lipids such as fat
body.

3. Guanidinium and lithium chloride. Guanidinium chloride has been used
for deproteinising ribosomes and at concentrations of 4 M is an effective
ribonuclease inhibitor [73, 74]. SDS can be used in a similar manner (see
below). Lithium chloride has been used to prepare RNA from the post-
mitochondrial supernatant as well as ribosomes [75, 76].

Tissue homogenisation

Homogenisation of the tissue and disruption of the cells can be achieved both by mechanical means and by the use of detergents.

1. Mechanical.   One of the most widely used techniques is the glass
homogeniser with a motor-driven teflon pestle (Potter-Elvejhem type). The
all-glass hand and Dounce homogenisers function in a similar way. These
provide a relatively gentle homogenisation of the tissue and are useful
when the isolation of long polysomes are required or when nuclear
breakage is to be minimized. When whole insects are to be homogenised
especially in the presence of deproteinisation agents, the Waring
blendor, Servall Omnimixer, or Polytron should be considered. We have
found the Polytron to be very satisfactory in the homogenisation of a
wide variety of insects.

2. Detergents.   The following non-anionic detergents are most frequently
used; Triton X-100; Tween 80; and Nonidet P40. They are usually used for
the rupture of cultured cells and have not been used extensively for
insect tissues.

Precipitation of RNA

        Once the RNA has been isolated in an aqueous phase it can be
precipitated by the addition of 2 volumes of 95% ethanol at $-20^{o}C$. The
efficiency of precipitation and the time required depend upon both the
concentration of RNA and cations. The presence of 0.2 $\underline{M}$ potassium
acetate at pH 6.0 has been found to be satisfactory for precipitation of
most RNAs. After 1-16 h the RNA is collected by centrifugation at
12,000 g for 20 min. If the original RNA concentration is less than
50 µg/ml the RNA precipitate can be more efficiently collected by
filtration through a Millipore filter. Excess ethanol is removed by a
stream of nitrogen and the RNA dissolved or eluted with distilled water
or an appropriate buffer.

VI   RIBOSOMAL RNA

        Although rRNA is the most abundant RNA in the cell it is somewhat
difficult to isolate free from some nuclease attack. Some of this may
occur naturally in the cell, and it has been found that the 26S rRNAs
from several insects contain a central break in the polynucleotide chain
[10, 12, 77]. The resultant two chains are held together by intra-
molecular hydrogen bonding and when subjected to a brief heat treatment or
agents that disrupt hydrogen bonds they dissociate to produce two

components of approximately equal size which form a single peak on a
sucrose gradient [10]. This cleavage may occur during the processing of
rRNA precursor. Because of this, if preservation of native molecular
weight is desired, conditions which maintain hydrogen bonding and
secondary structure must be maintained during the extraction. Inclusion
of NaCl or sodium acetate to a final concentration of 0.10-0.15 $\underline{M}$ is
important and all procedures should be performed between 0°-4^oC. After
the isolation of the rRNA the quality of the product is usually determined
by electrophoretic or sucrose gradient analysis as described below.

Isolation from whole cells

        This method is a modification of the procedure described by
Greenberg et al., [78] that we have used to prepare rRNA from the fat
body of Locusta migratoria. The fat body is homogenised in 10 vol of
0.02 $\underline{M}$ Tris-HCl, pH 7.4, 0.1 $\underline{M}$ NaCl, 0.5% (w/v) SDS and 0.1% (w/v)
bentonite. Buffers at pH 5 should function equally well and might reduce
ribonuclease activity. The homogenate is transferred to a beaker and an
equal volume of phenol saturated with 0.1 $\underline{M}$ NaCl is added. This is
stirred vigorously at 4^oC for 20 min and the resulting emulsion centrifuged
at 8,000 g for 10 min. The upper aqueous phase is carefully collected
and the phenol deproteinisation step is repeated upon it three times.
The final aqueous phase is collected and the RNA precipitated and
collected by centrifugation. At this point the precipitate contains rRNA,
tRNA, some mRNA and DNA as well as carbohydrate contaminants such as
glycogen. The low molecular weight RNAs, the carbohydrates and to some
extent the DNA can be removed by washing the pellet 3 times with 3 $\underline{M}$
sodium acetate [79, 80]. The pellet can then be dissolved in 0.01 $\underline{M}$
sodium acetate and treated with DNase by the method of Brown and Suzuki
[81].

Isolation from microsomes

        We have modified the procedure described for the isolation of
microsomes from rat liver [82]. Locust fat bodies are homogenised in
2.5 volumes of 0.05 $\underline{M}$ Tris-HCl, pH 7.8, 0.24 $\underline{M}$ KCl, 0.01 $\underline{M}$ MgSO$_4$, 0.01 $\underline{M}$
2-mercaptoethanol and 0.25 $\underline{M}$ sucrose with 8-10 strokes of a Dounce
homogeniser. The homogenate is centrifuged at 12,000 g for 30 min to
remove the nuclei, mitochondria and cellular debris and a fatty layer
which rises to the top. The supernatant is carefully removed and the
microsomes pelleted by centrifugation at 100,000 g for 70 min. The

microsomal pellets are resuspended in 3 ml of 0.01 $\underline{M}$ Tris-HCl, pH 7.4,
0.15 $\underline{M}$ NaCl. An equal volume of 3% SDS is added and the suspension
stirred for 10 min at room temperature. An equal volume of 88% phenol is
added and the mixture stirred for 20 min at 4°C. The 2 phases are
separated by centrifugation, the upper aqueous phase collected and the
phenol deproteinisation repeated upon it three times. The RNA in the
final aqueous phase is precipitated with 2 volumes of ethanol. At each
re-extraction of the aqueous phase with phenol a small percentage of the
RNA is lost but can be recovered by re-extraction of the phenol phases
with an equal volume of buffer. The RNA can be further purified as
described above.

Fractionation

        The ribosomal RNAs are conveniently separated according to their
sedimentaiton velocity by centrifugation through sucrose gradients [83]
or by polyacrylamide gel electrophoresis [84].

Sucrose gradients

        Linear gradients in the range of 5 to 30% are most commonly used
[83]. The sucrose solutions are prepared in buffers designed to maintain
the secondary structure of the RNA. It is advisable to add EDTA to ensure
the absence of divalent cations which can cause intermolecular aggregation.
The RNA sample is dissolved in the same buffer as the sucrose and
carefully layered on the top of the gradient. Approximately 0.5-1 mg of
RNA can be fractionated on a 12 ml gradient (see Fig. 1).

Polyacrylamide Gel Electrophoresis

        Electrophoretic separations of RNAs can achieve a superior
resolution to sucrose gradients but are usually less convenient on a
preparative scale. The larger ribosomal RNAs are separated on 2-3% gels
while lower molecular weight RNAs are separated on 5-10% gels. Gels of
less than 2.4% acrylamide can be physically strengthened by addition of
0.5% agarose [84].

VII MESSENGER RNA

        Despite the fact that mRNA comprises only a minor percentage of the
total RNA, newer techniques, especially those taking advantage of the
poly(A) found on the 3' end of many eukaryotic mRNAs have allowed

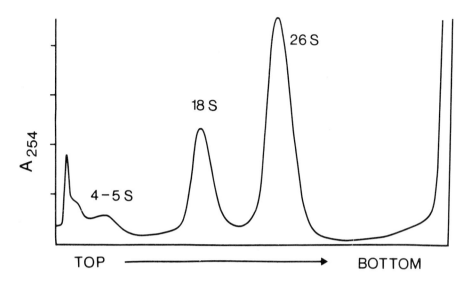

Figure 1. Sucrose gradient analysis of polysomal RNA (see below) from
female Locusta migratoria fat bodies. Polysomal RNA was layered onto a
12 ml linear gradient of 5-20% (w/v) sucrose in 0.03 M sodium acetate,
pH 5.0, 5 mM EDTA, 0.5% SDS and centrifuged at 283,000 g (max) for 6 h at
20°C. The gradient was analysed by continuous UV monitoring at 254 nm
[85].

purification of these molecules. These techniques use oligo(dT)-cellulose
[18, 86-88] and poly(U)-Sepharose [89-91] columns to which the poly(A) of
the mRNA hydrogen bonds. Cellulose itself has an affinity for poly(A),
as has nitrocellulose, and Millipore filters have been used to isolate
poly(A)-containing mRNA fractions [8, 92]. The length of the poly(A)
fragment is crucial to the precise behaviour of the mRNAs in these
procedures. The shorter the poly(A), the weaker is the affinity to the
pyrimidine polymer,and this results in increased difficulty in separating
the mRNA from the other RNA molecules. These procedures merely separate
poly(A)-containing RNAs from others; in order to isolate specific RNAs it
is common to use tissues that predominantly make one type of protein. The
isolation of fibroin mRNA from the posterior silk gland of B. mori [1] is
an example of this approach in insect tissues.

    One approach to the isolation of pure mRNAs involves the specificity
of the immunoprecipitation of nascent polypeptide chains on polysomes.
This approach has not yet been successfully applied to an insect system.

    Although the presence of poly(A) is useful in the separation of the

mRNA from other RNAs it can cause problems in the deproteinisation step.
It has been shown that mRNA is selectively lost during phenol extraction
because of binding of the poly(A) to the denatured proteins [95]. This
binding is enhanced by monovalent cations, neutral pH and low
temperature. The mRNA can be recovered from the interphase by re-extraction with buffer at pH 9 and low monovalent cation content [94, 95].
Similar results can be achieved by extraction at 60°C [67, 96], but this
may lead to some degradation of the mRNA. The presence of chloroform
during the phenol deproteinisation (phenol:chloroform, 1:1) also prevents
the association of the poly(A) with denatured proteins [81, 97, 98].

We have used the extraction of RNA in the presence of phenol at
pH 7.4 followed by the re-extraction of the interphase at pH 9 for
isolation of mRNA from locust fat body [94] and the chloroform:phenol
mixture for isolation of total RNA from whole Drosophila and locust fat
body.

Isolation from whole cells

Locust fat body - pH 9.0 RNA

The fat body is homogenised in 20 vols (wet w/v) 0.05 $\underline{M}$ Tris-HCl,
pH 7.4, 0.1 $\underline{M}$ NaCl, 0.5% SDS and 0.1% bentonite. To this is added an
equal volume of 88% phenol and the mixture is shaken for 30 min at 4°C
and then centrifuged at 8,000 g for 10 min. The aqueous phase is removed
and is designated pH 7 RNA. To the interphase and phenol phase is added
an equal volume of 0.1 $\underline{M}$ Tris-HCl, pH 9.0, 0.5% SDS. This is shaken for
15 min and centrifuged. The aqueous phase is removed and the interphase
and phenol are re-extracted with an equal vol of the pH 9.0 buffer. Both
aqueous phases are pooled and mixed with an equal volume of phenol
saturated with 0.1 $\underline{M}$ Tris-HCl, pH 9.0. This is shaken, centrifuged and
this deproteinisation step repeated on the resultant aqueous phase. The
RNA is precipitated, centrifuged, and the pellet washed three times with
3 $\underline{M}$ sodium acetate, pH 6.0. The RNA precipitate is then treated by
DNase under the conditions described by Brown and Suzuki [81].

Whole Drosophila - SDS-phenol-chloroform

This technique is a combination of that described by Palmiter [46]
for chick oviduct and that used by Spradling et al., [99] for Drosophila
cells. To each gram of flies is added 10 ml of 0.1 $\underline{M}$ sodium acetate,
pH 5.0, 0.025 $\underline{M}$ NaCl, 5 m$\underline{M}$ MgCl$_2$, 0.5% SDS containing 25 µg/ml polyvinyl
sulfate   35 µg/ml spermine and 10 ml of buffer-saturated phenol. This

is homogenised for 2 min at 4°C with the Polytron and centrifuged at
8,000 g for 10 min. The mixture is then poured into shaker tubes and
shaken for 10 min at 4°C. Chloroform (10 ml) is added and the mixture is
again shaken for 10 min at room temperature. The mixture is centrifuged
and the organic phase removed by aspiration, leaving behind the aqueous
phase and the flocculent interphase. To this is added 20 ml chloroform
and the mixture shaken and centrifuged. The chloroform is removed and
the chloroform extraction repeated until the interphase disappears. The
upper phase is removed and the RNA precipitated with 2 vols of ethanol at
-20°C. The RNA precipitate is collected by centrifugation and washed
three times with 3 $\underline{M}$ sodium acetate, pH 6.0. The RNA can be stored as a
precipitate under ethanol at -20°C or dissolved in 0.1 $\underline{M}$ sodium acetate
pH 7.0, reprecipitated, dried with N_2, dissolved in distilled water or a
dilute buffer and stored frozen in aliquots.

Isolation from polysomes
        Polysomes can occur free in the cytoplasm or bound to the membranes
of the endoplasmic reticulum. The classical procedure for the isolation
of polysomes from animal cells is the treatment of the post-mitochondrial
supernatant with detergent to release the membrane bound polysomes. The
ratio of bound to free polysomes will chiefly depend on whether the tissue
in question is making a large amount of protein for export. The procedure
to be outlined has been used for a tissue that is primarily involved in
synthesizing proteins for export, namely the fat body of the adult female
locust. The procedure is modified from that used to isolate polysomes
from chick oviduct [100]. The fat body is dissected and rinsed in 0.85%
(w/v) NaCl containing 0.1% bentonite, blotted on filter paper and
immediately transferred to 10 volumes (v/wet w) 0.02 $\underline{M}$ Tris-HCl pH 7.4,
0.025 $\underline{M}$ KCl, 0.01 $\underline{M}$ MgSO$_4$, 0.25 $\underline{M}$ sucrose, 0.5 mg/ml sodium heparin and
0.1% (w/v) bentonite at 4°C. Homogenisation is achieved by 5-7 strokes
with a Dounce homogeniser with a loose pestle at 4°C. Triton X-100 and
sodium deoxycholate are added to a final concentration of 1% each and
homogenisation is continued with a further 3 strokes. The viscosity of
the homogenate increases because of lysis of the nuclei and in this state
it is passed through a syringe fitted with four layers of cheesecloth to
remove fragments of tracheae, cellular debris and some fat.
        The filtrate is centrifuged on a sucrose gradient as described in
Fig. 2 and after centrifugation the lipid layer present at the top of the

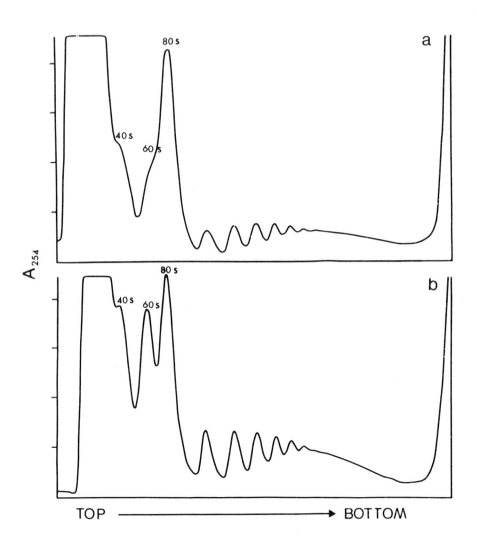

Fig. 2.   Sucrose gradient analysis of polysomes from (a) male and (b)
female <u>Locusta migratoria</u> fat bodies.   The extract was layered onto a
12 ml linear gradient of 15-50% (w/v) sucrose in 0.02 $\underline{M}$ Tris-HCl, pH 7.4,
0.025 $\underline{M}$ KCl and 0.01 $\underline{M}$ MgSO$_4$ containing 80 µg/ml sodium heparin.   This
was centrifuged at 283,000 $\underline{g}$ (max) for 2 h at 6°C.   The gradient was
analysed by continuous monitoring by A$_{254}$ [85].

gradient is removed so that it does not interfere with collection. The
gradients are displaced upwards with a 60% (w/v) sucrose solution containing 20% w/v sodium bromide to increase the density and 0.1% (w/v) potassium
acid phthalate to indicate the end of the gradient by UV absorption. The
A_{254} of the gradient is continually monitored with a UV analyser (Fig. 2).

    In order to confirm that the polysomes are not merely aggregates
of ribosomes and to determine their length, they can be directly
visualized with the electron microscope using the technique described by
Vignais et al., [101]. Three to four drops of the sucrose gradient
fraction are placed on a carbon-formvar coated grid. Special care is
taken to remove the excess sucrose solution without completely drying the
grid. One drop of 2% (w/v) uranyl acetate is immediately added, and after
15-30 seconds the excess is removed and the grid dried at room temperature
for 2-3 h (Fig. 3).

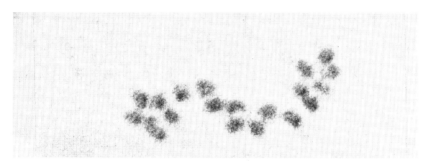

Fig. 3. Electron micrograph of a polysome from female Locusta migratoria
fat body [102].

    For the isolation of polysomal RNA, fractions of the sucrose
gradient corresponding to polysomes (see above) are pooled and precipitated
with 2 vols of 95% ethanol at -20°C. The precipitate is collected by
centrifugation and dissolved in 0.02 M sodium acetate, pH 5.0, 5 mM EDTA
containing 0.5% SDS. This is layered on a 12 ml linear gradient of 5-20%
(w/v) sucrose in the same buffer. This is centrifuged at 283,000 g (max)
for 6 h at 20°C and fractionated with continuous UV monitoring. A
sedimentation profile of female locust polysomal RNA is shown in Fig. 1.
Fractions from such a separation of polysomal RNA can be assayed for mRNA
activity as described below.

Further Purification of mRNAs

        There are two basic approaches to the further purification of
mRNAs. When the mRNA in question has a distinctive size and is present
in relatively high concentrations in a tissue, sucrose gradient
sedimentation as described above can be very successful. Three such steps
are sufficient to purify the very large silk fibroin mRNA from B. mori
[81]. A similar approach was used for moth chorion protein mRNAs [103,
104] and this was coupled with labeling procedures and polyacrylamide gel
electrophoresis for an analysis of the kinetics of synthesis of the mRNAs.
Even if the mRNA to be isolated is not the predominant one in the tissue,
sucrose gradient sedimentation can be used to enrich for it. The other
major approach is to take advantage of the poly(A) tail found on most
mRNAs and separate these molecules from the poly(A) minus RNAs such as
rRNAs. This approach utilizes the affinity of the poly(A) for cellulose
and nitrocellulose or hybridisation of the poly(A) to oligo(dT)-cellulose
and poly(U)-Sepharose. Poly(U)-Sepharose usually gives cleaner
preparations than oligo(dT)-cellulose but is subject to ribonuclease
hydrolysis and is therefore less stable. Hybridisation to oligo(dT)-
cellulose columns coupled with sucrose gradient centrifugation and
electrophoresis were used to purify moth chorion mRNAs that were
ultimately employed for the making of complementary DNA sequences [104,
105]. Another very elegant study on the processing of fibroin mRNA used
a more selective affinity chromatographic approach by employing
Sephadex-bound polynucleotides complementary to fibroin mRNA sequences
[93].

        We have been using oligo(dT)-cellulose columns for purification of
the mRNA for vitellogenins from total RNA extracted from the fat body of
L. migratoria and whole Drosophila. The method used is essentially the
procedure of Aviv and Leder [86] as modified by Spradling et al., [36]
and is described in Fig. 4. There is a marked increase in specific
messenger activity following this procedure (Table 1).

        Several passages through oligo(dT)-cellulose columns are usually
required to remove all the pol-(A) minus RNA. The poly(A) plus mRNA can
be fractionated on sucrose gradients or polyacrylamide gels and assayed
for increase in specific activity over total RNA in a cell-free protein
synthesizing system.

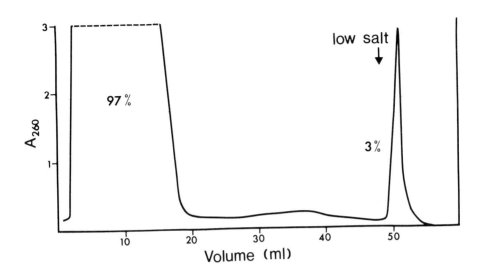

Figure 4. Oligo(dT)-cellulose chromatography of female adult <u>Drosophila</u>
RNA. Total RNA was extracted by the SDS-phenol-chloroform procedure and
(100 A$_{260}$ units) dissolved in 0.01 <u>M</u> Tris-HCl, pH 7.4, 0.4 <u>M</u> NaCl, 1 m<u>M</u>
EDTA and 0.5% SDS. This was applied to an oligo(dT)-cellulose
(Collaborative Research T-2, Waltham, Massachusetts) column (0.6 x 3 cm)
pre-equilibrated with the same buffer and washed with the buffer. The
poly(A) containing mRNA was eluted with 0.01 <u>M</u> Tris-HCl, pH 7.4, 0.1% SDS
(low salt) [106].

<u>Cell-free protein synthesis</u>

        In order to monitor the isolation and purification of biologically

active mRNA a cell-free protein synthesizing system is usually used. There

are a number of such systems available; reticulocyte lysate [107, 108];

Krebs II ascites tumor [109, 110]; wheat embryo [111-113]; and others [114,

115]. An alternative assay system of high sensitivity is provided by

<u>Xenopus</u> oocytes [116]. We have found the wheat embryo system as described

by Roberts and Patterson [111] to be satisfactory for the translation of

insect mRNAs. The preincubated wheat germ extract is prepared as described

by Roberts and Patterson except the wheat germ is obtained from General

Mills, Vallejo, California. The cell-free protein synthesizing assay

(50 μl) contains [115]: 15 μl of wheat germ extract; 20 m<u>M</u> HEPES, pH 7.6;

2 m<u>M</u> dithiothreitol; 1 m<u>M</u> ATP; 220 μ<u>M</u> GTP; 8 m<u>M</u> creatine phosphate; 8 μg/ml

creatine phosphokinase; 19 **unlabeled amino acids** (minus leu) 20-30 μ<u>M</u> each

and 1μCi of ^{3}H-leucine (50 Ci/mMole). The optimum level of magnesium
acetate, KCl and the RNA preparation should be determined for each wheat
germ extract and for each RNA preparation. The magnesium acetate optimum
is usually less than 5 m$\underline{M}$ added and about 80 m$\underline{M}$ added KCl.

Table 1 shows the requirements of the system and the stimulation
by various locust and Drosophila RNA preparations. The reaction mixture is
incubated at 25°C for 90 min and the total amount of protein synthesized
estimated in one of two ways: (1) five μl aliquots are pipetted onto 2.5 cm
Whatman 3 MM paper discs, dried and placed into 10% trichloroacetic acid
(TCA) for 20 min. The filters are then heated at 90°C in 5% TCA for 20 min
to break down the aminoacyl-tRNA, washed for three 20 min periods in 5% TCA,
washed with ethanol:ether (3:1 v/v), dried and counted in a scintillation
counter; (2) aliquots are pipetted into 4 ml of 10% TCA and 0.1 mg of
carrier protein is added. This is allowed to stand at 0°C for 10 min and
the precipitate centrifuged. The precipitate is resuspended in 5% TCA and
again centrifuged. This is repeated 2 times, then the precipitate suspended
in 5% TCA is heated to 90°C for 20 min. The precipitate is collected on
Millipore filters which are thoroughly dried and counted in a scintillation
counter.

The Whatman 3 MM procedure consistently yields higher cpm retained
than the Millipore filtration technique. The reason for this is not clear
but because the Millipore filtration more closely reflects the quantity of
protein that is recoverable for analysis we have usually utilized this
procedure. It can be directly compared to antibody precipitation of in
vitro synthesized protein which is also recovered by Millipore filtration.
The antibody precipitation technique is a very powerful tool for the
analysis of mRNAs for specific proteins. Comparison of the total TCA
precipitated cpm and the specific antibody precipitated cpm provides an
estimate of the percentage of the mRNA for a specific protein.

The product synthesized in the cell-free protein synthesizing system
can be analysed by SDS-electrophoresis [118]. This analysis can be
performed on the total TCA precipitable material or better on that material
specifically precipitated by an antibody made against the product under
study. We have found that RNA preparations from Drosophila are capable of
directing the synthesis in the wheat germ system of polypeptides of
molecular weight larger than 50,000. RNA preparations devoid of mRNA
activity can stimulate the endogenous activity of the wheat germ system up
to three fold [119] and so care must be taken in establishing that the

Table 1.   Stimulation of incorporation of ^{3}H-leucine into TCA precipitable
material in a wheat-germ cell-free protein synthesizing system as directed
by insect RNAs.

Source of RNA	µg RNA	cpm/5 µl reaction* Mixture	cpm/ µg RNA
Control	–	1,652	–
Rabbit globin mRNA	1	40,626	390,630
Locust female	10	11,790	10,247
Locust male	20	15,683	7,070
Drosophila female	9	14,178	14,226
Drosophila male	5.6	10,805	16,541
Drosophila female poly(A) plus	2.7	32,629	117,042
Locust female and ATA (50 µM)	10	1,432	–

* The assay mixture (50 µl) was incubated at 25°C for 90 min and 5 µl
aliquots precipitated with TCA, washed, collected on Millipore filters and
counted.   Total RNA was prepared from whole Drosophila and locust fat
bodies by the SDS-phenol-chloroform technique.   Poly(A) plus RNA was
prepared as described under Fig. 4.   ATA (aurintricarboxylic acid) is an
inhibitor of the initiation of protein synthesis.   Values represent the
average of three assays [117].

polypeptides being synthesized are being directed by the added RNA.

VIII TRANSFER RNA

        The preparation and analysis of tRNAs from several insects [3-5,
120,123] has proved to be relatively simple and there is every reason to
believe that the techniques can be easily adapted to any insect tissue.
This section deals with the isolation of tRNA and aminoacyl-tRNA synthetases,
the chromatographic examination of the isoaccepting tRNA species,
purification of tRNAs, the minor nucleoside analysis and the codon
recognition properties of the individual species.

Isolation of total tRNA

        Although there may be a high ribonuclease activity in many insect
tissues (especially Dipteran) the kinds of stringent precautions that are
essential for mRNA and rRNA are not necessary for tRNA.   The most
convenient technique involves a combination of the phenol extraction
technique of Kirby [61] and DEAE-cellulose step of Kelmers et al., [124].

To every gram (wet weight) of insect tissues is added 5 ml of 88%
phenol and 5 ml of buffer A containing 0.1 $\underline{M}$ NaCl, 0.01 $\underline{M}$ MgCl$_2$, 1 m$\underline{M}$
2-mercaptoethanol and 0.01 $\underline{M}$ Tris-HCl, pH 7.5. The tissue is homogenised
(Waring blendor, Omnimixer, Virtis and Polytron perform equally well) for
1 min at full speed while keeping the homogenising vessel at 0-4°C. The
material is cooled for 5 min and the 1 min homogenisation repeated. The
homogenate is centrifuged at 8,000 g for 10 min and the upper aqueous
layer carefully removed. The aqueous layer is kept on ice and the phenol
re-extracted with an equal volume of buffer A. To the pooled aqueous
phases is added an equal volume of phenol and the deproteinisation procedure
repeated. The RNA is precipitated with ethanol.

A DEAE-cellulose column is prepared and equilibrated with buffer A.
As a rule of thumb 1 g of insect tissue yields approximately 10 mg of total
RNA and 1 ml of DEAE-cellulose (70 mg dry powder) binds approximately 3 mg
of RNA. Therefore the minimum size of the DEAE-cellulose column should be
about 4 ml for every gram of starting insect tissue. The RNA is dissolved
in buffer A and applied to the DEAE-cellulose column and the column is
washed with buffer A containing 0.3 $\underline{M}$ NaCl. This removes the uncharged
glycogen and the very small nucleic acid fragments. When the A_{260} reading
(1 cm path length) of the eluate has stabilised to about 0.1 the tRNA can
be eluted with buffer A containing 1.0 $\underline{M}$ NaCl. The peak of A_{260} absorbing
material eluted is pooled and precipitated with ethanol. The precipitate
is collected and dissolved in distilled water, extensively diaylsed against
distilled water and freeze dried. This dried material can be stored
indefinitely in the refrigerator.

Recently we have found a variation of this procedure gives more
reliable results. The technique is essentially the same except the 0.01 $\underline{M}$
Tris-HCl buffer used in the extraction and column steps is replaced with
0.01 $\underline{M}$ sodium acetate, pH 4.5. This lower pH reduces nuclease activity
and consistently yields tRNA with a higher biological activity. The pooled
aqueous phases can also be directly applied to the DEAE-cellulose column
without precipitation. The column requires a somewhat longer washing
period but again usually yields a better product. As these procedures are
performed at pH 4.5 the ester bond between the amino acid and tRNA is
stabilised and aminoacyl-tRNA may be recovered. The amino acids can be
stripped from the tRNA by incubation of the tRNA in 1.8 $\underline{M}$ Tris-HCl, pH 8.0
for 2 h at room temperature.

An analysis of such a tRNA preparation from Drosophila by Sephadex
G-100 chromatography is shown in Figure 5. This shows that the material

comprises about 50% 4S material, 9% 5S rRNA and 41% high molecular weight
RNA that is probably derived from breakdown of the ribosomal RNA.

Preparation of aminoacyl-tRNA synthetases

        The main object in this procedure is to isolate an enzyme
preparation that contains the activity of all 20 aminoacyl-tRNA synthetases
and that is free of tRNA. The most convenient procedure is a modification
of the method of Muench and Berg [125].

        To every gram of insect tissue is added 4 ml of buffer B containing
0.01 $\underline{M}$ Tris-HCl, pH 7.5, 0.01 $\underline{M}$ magnesium acetate, 0.01 $\underline{M}$ 2-mercaptoethanol,
1 m$\underline{M}$ phenylmethylsulfonyl fluoride, 1 m$\underline{M}$ phenylthiourea and 10% glycerol

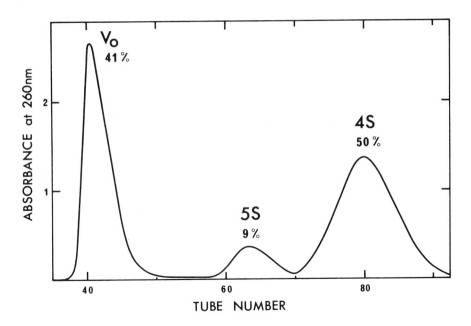

Fig. 5.   Chromatography of a <u>Drosophila</u> tRNA preparation on a Sephadex
G-100 column.   The crude tRNA (300 A_{260} units) was applied to a 195 x 2.5
cm Sephadex G-100 column, and eluted with 0.01 $\underline{M}$ Tris-HCl, pH 7.5, 0.01 $\underline{M}$
magnesium acetate and 0.05 $\underline{M}$ NaCl at 6 ml/h with 10 ml fractions being
collected [126].

and the tissue is then homogenised with the Polytron. The resultant
homogenate is centrifuged at 105,000 g for 90 min. The supernatant is
carefully removed as the ribosomal pellet is usually high in ribonuclease
activity. The NaCl concentration of the supernatant is adjusted to 0.3 $\underline{M}$

and applied to a DEAE-cellulose column equilibrated with buffer B containing
0.3 $\underline{M}$ NaCl. The supernatant contains approximately 1 mg of tRNA for every
gram of starting tissue. The minimum size of DEAE-cellulose column used
should be approximately 1 ml for every 3 g of starting tissue. The A_{280}
absorbing material eluted from the column is collected and pooled. This
material is dialysed against buffer B containing 50% glycerol overnight
with at least two changes of the buffer. This procedure concentrates the
preparation approximately three fold. The preparation can be stored at
-70°C in small aliquots for over a year without significant loss of
biological activity.

Aminoacylation of the tRNA

        In order to obtain a true qualitative and quantitative analysis of
the tRNAs the assay conditions for the aminoacyl-tRNA synthetases must be
carefully established. The most important factors to take into consideration are the Mg^{++}:ATP ratio and the concentration of the amino acid.
Factors such as pH optimum and monovalent cation requirements are usually
not quite as important.

        A generalised assay which has proved successful for many insect
aminoacyl-tRNA synthetases is as follows; 50 m$\underline{M}$ Tris-HCl, pH 7.5, 10 m$\underline{M}$
Mg^{++} , 4 m$\underline{M}$ ATP, 30 m$\underline{M}$ KCl, 1 mg/ml crude aminoacyl-tRNA synthetases,
1-0.5 mg/ml crude tRNA, 25-50 μM labeled amino acid, 50 μM each unlabeled
amino acids, 4 m$\underline{M}$ 2-mercaptoethanol (Table 2). Each aminoacyl-tRNA

Table 2 Generalised incubation mixture for tRNA aminoacylation

	+ tRNA	− tRNA control
Premix*	50 µl	50 µl
tRNA (in 0.01 $\underline{M}$ Tris-HCl, pH 7.5, 5 m$\underline{M}$ magnesium acetate, 1 m$\underline{M}$ 2-mercaptoethanol)	25 µl	−
buffer (tRNA dissolving)	−	25 µl
labeled amino acid (neutralised)	25 µl	25 µl
aminoacyl-tRNA synthetase	75 µl	75 µl
19 unlabeled amino acids	10 µl	10 µl
water	15 µl	15 µl
	200 µl	200 µl

* Premix contains the appropriate concentrations of Mg^{++}, ATP, KCl, and
Tris-HCl.

synthetase will require slightly different conditions [126, 127] but this
represents a good starting point.

All the components of the assay are added, except the tRNA and enzyme, to
the bottom of small tubes in an ice bath. At this time the glycerol and
any low molecular weight material remaining in the aminoacyl-tRNA
synthetase preparation must be removed. This is conveniently achieved by
passing 1 ml of the extract through a 1.5 x 10 cm Sephadex G-25 column.
The column is equilibrated and eluted with a buffer containing 0.01 $\underline{M}$
Tris-HCl, pH 7.5, 5 m$\underline{M}$ magensium acetate, 0.01 $\underline{M}$ 2-mercaptoethanol, 1 m$\underline{M}$
phenylmethylsulfonyl fluoride and 1 m$\underline{M}$ phenylthiourea. One ml fractions
are collected and the fraction with the highest A_{280} reading is used as
the final enzyme source. This procedure dilutes the preparation 2-3 fold.

The tRNA is added and the reaction initiated with the enzyme and
vortexing. The reaction can be carried out at various temperatures up to
37°C although lower temperatures (22°C) minimise the ribonuclease effect
on the tRNA. The kinetics of the reaction should be followed and this is
conveniently achieved by pipeting 25 µl or 50 µl aliquots of the reaction
mixture onto Whatman 3 MM filter paper discs (2.5 cm) and removing
acid-soluble radioactive amino acids [128]. The discs are labeled with
pencil. Each sample is spread over the disc which is briefly air dried
and placed in a beaker of ice-cold, 10% TCA (10 ml for every filter). The
filters are washed for 20 min with occasional stirring and then the 10%
TCA replaced with cold 66% ethanol containing 0.5 $\underline{M}$ NaCl. Washing is
continued for 20 min and then this is replaced with cold 5% TCA. The
filters are finally rinsed in a mixture of ethanol ether (3:1) and then
in ether. The filters are dried under a heat lamp and counted in a
scintillation counter.

A plateau level of charging should be reached within 10 min to
ensure reasonably complete aminoacylation of the tRNA. The rate of
aminoacylation with varying amounts of the Tenebrio molitor leucyl-tRNA
synthetase is shown in Fig. 6 [123, 127, 129]. At reduced enzyme levels
both the rate and extent of aminoacylation are markedly reduced. At these
limiting enzyme levels an apparent plateau is reached but this clearly
does not represent the amount of leucine tRNA in the preparation. Further
complications can arise when the different isoaccepting species are
aminoacylated at different rates. This will be discussed later when the
isoacceptors are analysed.

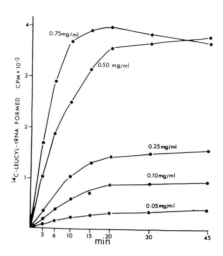

Fig. 6. Effect of Tenebrio leucyl-tRNA synthetase concentration on the rates and extent of ¹⁴C-leucyl-tRNA formation. [127]

If satisfactory aminoacylation is not achieved, various parameters such as enzyme, Mg⁺⁺ and ATP concentrations should be varied. The best way of examining these parameters is by initial rates of reaction rather than the final extent of aminoacylation. The reaction tubes are set up as before but are terminated after 2-4 min by addition of 3 ml 10% TCA containing a 100 fold excess of non-radioactive amino acid. After standing 10 min the precipitate is filtered through a Millipore filter under vacuum. The filter is washed with 3 x 5 ml of 5% TCA and transferred to a larger support and washed with more 5% TCA to ensure washing of the edges of the filter. If a high background control (no tRNA) is obtained even after this procedure, the initial precipitate should be centrifuged and resuspended in 5% TCA prior to filtering.

The effect varying Mg⁺⁺ and ATP concentrations on the initial rates of the Tenebrio leucyl-tRNA synthetase [127, 129] is shown in Fig. 7. This enzyme in the crude state at least has an unusual Mg⁺⁺:ATP optimum of 0.63.

When satisfactory aminoacylation conditions have been established the amount of tRNA for a given amino acid in the crude tRNA preparation can be established. This is usually expressed as pmoles of amino acid accepted per A_{260} unit of tRNA and is calculated as follows:

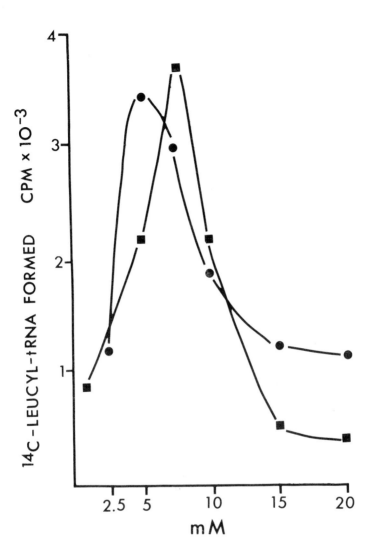

Fig. 7. Effect of ATP and $MgCl_2$ concentration on the initial rate of
Tenebrio ^{14}C-leucyl-tRNA formation. Reactions were terminated after 2 min
with 10% TCA and the precipitate collected on Millipore filters; ■—■ ATP
at 5 mM $MgCl_2$; ●—● $MgCl_2$ at 8 mM ATP [127]

$$\frac{\text{cpm of a.a. accepted}}{\text{spec. act. of a.a. (pCi/pmole)}} \quad \text{x} \quad \frac{1}{2.22} \quad \text{x} \quad \frac{100}{\text{efficiency of scintillation counter}}$$

$$\text{x} \quad \frac{1}{A_{260} \text{ units tRNA}} \quad = \text{x pmoles/A}_{260}$$

Table 3. Acceptance of amino acids by Drosophila tRNA*

^{14}C-labeled amino acid	^{14}C-labeled amino acid accepted (pmole) per A_{260} unit of tRNA		
	Time (min)		
	10	20	30
Alanine	45	42	42
Arginine	45	42	41
Asparagine	37	37	37
Aspartic acid	30	38	39
Cysteine	17	17	16
Glutamic acid	23	32	32
Glutamine	27	35	39
Glycine	72	71	71
Histidine	36	36	36
Isoleucine	32	32	32
Leucine	35	37	38
Lysine	67	60	60
Methionine	45	49	47
Phenylalanine	34	35	34
Proline	31	32	33
Serine	69	71	70
Threonine	36	35	33
Tryptophan	28	27	29
Tyrosine	20	20	20
Valine	51	56	54

*
Approximate total maximum acceptance, 828 pmole/A_{260} unit [126].

The acceptance of all 20 amino acids by adult Drosophila tRNA [126] is shown in Table 3. Note that not all of the amino acids reach plateau levels by 10 min and this is because better aminoacylating conditions could not be easily established. The total acceptance for all 20 amino acids per A_{260} unit is about 828 pmoles, which is in good agreement with the 50% content of 4S material in these preparations (Fig. 5). The theoretical acceptance for a pure tRNA is approximately 2000 pmoles/A_{260} unit.

These assays are very reproducible and can be used to study
quantitative changes during development. A good standard curve can be
obtained by varying the amount of tRNA in the preaparation (Fig. 8). The
most variable factor is the degree of A_{260} contamination of the 4S material
especially when tRNAs from different developmental stages are compared.
A better comparison is between the acceptance of one amino acid to that of
another [121, 123].

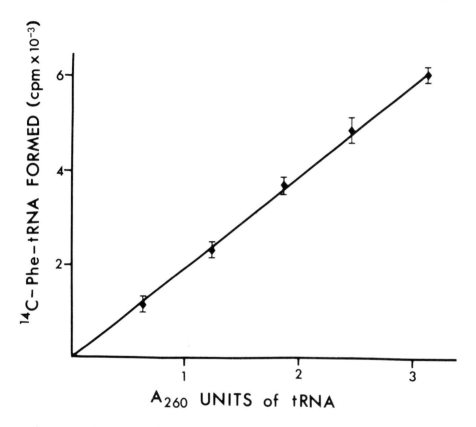

Fig. 8.   Acceptance of ^{14}C-phenylalanine with increasing amounts of
Drosophila   tRNA [130].

Analysis of isoaccepting tRNA species

Two basic approaches may be used to determine the number and
relative amounts of the isoaccepting species. Small analytical chromatographic columns can be used to separate pre-labeled aminoacyl-tRNAs. This
procedure requires relatively little tRNA. The alternative approach which

requires greater amounts of tRNA is to chromatograph tRNA on larger columns and assay the fractions for amino acid acceptor activity.

Analysis of labeled aminoacyl-tRNAs on analytical columns

In this procedure the tRNA is charged with a labeled amino acid, purified from the reaction mixture and chromatographed on the analytical column.

Preparation of labeled aminoacyl-tRNA

The reaction mixture as described previously can be used or increased proportionately such that 20,000-40,000 cpm of labeled aminoacyl-tRNA is prepared. The aminoacyl-tRNA is isolated from the components of the reaction mixture by DEAE-cellulose chromatography [131]. Small DEAE-cellulose columns (0.7 x 4 cm) are equilibrated at 4°C with buffer C containing 0.25 $\underline{M}$ NaCl, 0.01 $\underline{M}$ MgCl$_2$, 0.01 $\underline{M}$ sodium acetate, pH 4.5 and 1 m$\underline{M}$ 2-mercaptoethanol. The aminoacylation mixture is applied directly to the column and washed with buffer C until the A$_{260}$ reading has stabilized to below 0.1. The aminoacyl-tRNA is eluted with buffer C containing 0.75 $\underline{M}$ NaCl and at this stage can be frozen and stored.

Analytical Chromatography

The most convenient system is reversed phase-5 [132] although BD-cellulose [133, 134] has certain advantages in some circumstances. The RPC-5 is commercially available or can be prepared from the basic components. The RPC-5 is suspended in buffer C containing 0.45 $\underline{M}$ NaCl and passed through a 200 mesh sieve to remove the clumps formed in the coating process. The slurry is degassed and poured into a 0.9 x 30 cm Chromatronix jacketed column partially filled with buffer C. The packing is compacted under maximum flow of a high pressure pump such as the Milton Roy Minipump to give a bed height of about 13 cm. If the material is not compacted peak widths are usually broad and resolution is poor. Once equilibrated with buffer C the column is ready to use.

The aminoacyl-tRNA stored frozen in buffer C containing 0.75 $\underline{M}$ NaCl should be diluted to below 0.45 $\underline{M}$ NaCl and loaded onto the column. The column must be maintained at a constant temperature. Generally 37°C gives sharpest peaks and best resolution but deacylation of certain aminoacyl-tRNAs is very rapid at these temperatures and then 22°C or 30°C should be used [120]. These columns are conveniently eluted by a 100 ml NaCl gradient between 0.5 and 0.7 $\underline{M}$ NaCl. The range and shape of the gradient will depend on the isoacceptors being resolved. Maximum resolution is

achieved with a 15 ml/h flow rate and collection of 200 fractions (0.5 ml).
The radioactivity in the fractions is determined by mixing with a water
miscible scintillation fluid and counting in a scintillation counter.

Using this procedure changes in aminoacyl-tRNA can be followed
through development and compared to changes in other isoaccepting species
(Fig. 9) [120, 121, 123]. From the results certain inferences about the
nature of the change can be made. In this example [121] the tRNAs
changing, recognised codons of the type XA_U^C so it seemed apparent that the
change involved the anticodon loop in some way. As there did not appear
to be a quantitative change in these tRNAs the most likely explanation was
that the two peaks were derived from the same gene (homogeneic) but were
chromatographically different because of the degree of post-transcriptional
modification. The analysis of modified bases in tRNAs will be discussed
later.

It is imperative that the aminoacylation be taken to completion in
order to obtain meaningful profiles. In certain instances the isoaccepting
species are aminoacylated at different rates and this can be lead to
artificial profiles. In the Tenebrio leucine tRNA system two isoacceptors
are aminoacylated poorly compared to the major one and under poor charging
conditions appear quantitatively reduced [123, 127, 129]. Such results
can lead to incorrect conclusions concerning the changes and roles of tRNAs
during development [135, 136].

Chromatography of tRNAs

In the analysis just described an isoaccepting profile could be
determined with few A_{260} units of tRNA, but when tRNA is chromatographed
followed by determination of acceptor activity by the various factions,
about 10-20 mg of tRNA are required [126]. The advantage of this approach
is that it can be directly applied to preparative scale procedures. Again
the most useful column procedures involve RPC-5, BD-cellulose and to a
certain extent DEAE-Sephadex and other chromatographic methods. The
chromatography of 1 g of tRNA from adult Drosophila on BD-cellulose is
shown in Fig. 10 [137]. The acceptance of ^{14}C-serine is fractionated into
several major peaks which are numbered according to their elution position
on RPC-5 columns.

The easiest way to assay the acceptance of an amino acid in column
fractions is the paper disc procedure of Cherayil et al., [138]. Aliquots
of 0.1 ml of the column fractions are applied to numbered filter paper
discs (Whatman 3 MM, 2.3 cm) which are stapled to a 8 x 125 cm strip of

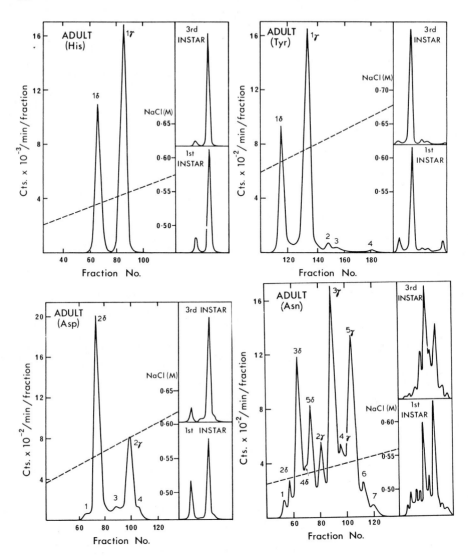

Fig. 9. Chromatography of ^{14}C-labeled histidyl-, tyrosyl- aspartyl- and asparaginyl-tRNAs from <u>Drosophila</u> first- and third-instar larvae and adults on RPC-5 columns. Asparaginyl- and histidyl-tRNAs were chromatographed at 22°C and tyrosyl- and aspartyl-tRNAs at 37°C. The labeled aminoacyl-tRNAs (100-500 µg total tRNA) were loaded onto a 13 cm x 0.9 cm column which was eluted at 15 ml/h using a 100 ml gradient of NaCl containing 0.01 $\underline{M}$ sodium acetate (pH 4.5), 0.01 $\underline{M}$ MgCl$_2$ and 1 mM 2-mercaptoethanol, with a fraction size of 0.5 ml. Radioactivity was determined in the fractions by addition of 5 vol. of Aquasol and counting in a scintillation counter [120].

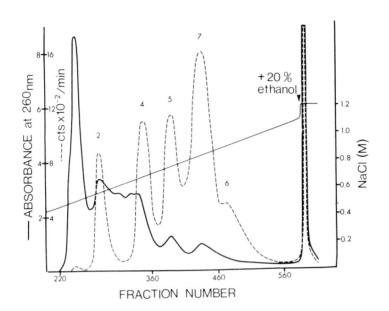

Fig. 10.   Separation of Drosophila serine tRNAs on a BD-cellulose column.
Drosophila tRNA (1g) was applied to the column (100 x 2.5 cm) in 0.35 M
NaCl and eluted by a 12-1 linear gradient from 0.35 to 1.1 M NaCl.   The
flow rate was 200 ml/h; 20 ml fractions were collected. ^{14}C-serine
acceptance was determined by the paper disc procedure [137].

40 mil ribbed polyethylene sheet (Handi-Mat, Fisher Scientific) [126].   The
paper discs are dried and the Handi-Mat strips threaded into a film-
developing reel.   The reel is immersed in a bath of 75% ethanol containing
0.03 M KCl at 0-4°C, and the paper discs are washed free of excess salt by
gentle mechanical rotation for 20 min.   The Handi-Mat strip is removed
from the reel, the discs are blotted dry and finally allowed to air dry.
The tRNA on the paper discs is then incubated at room temperature, with
0.1 ml of the incubation mixture containing the radioactive amino acid
activating enzymes etc., in a humidified styrofoam box.   After incubation
for 30 min the paper discs are dried briefly until the surface no longer
has a wet sheen, threaded back into the developing reel and immersed in a
bath of 10% TCA.   The filters are washed with the same solutions described
under "Aminoacylation of tRNA".

     This procedure has several advantages over the normal test tube
assay.   By removing all the salt from the tRNA aliquots, there is no
problem with differential extents of charging due to salt inhibition.   The

washing procedures are faster and little glassware is used. However, a
higher concentration of enzyme is required for this procedure.

Purification of tRNA species

There are two basic approaches to purifying tRNAs. The first is a
continuation of that in the previous section, where a tRNA is isolated by
several different chromatographic procedures. The second involves
modification of certain tRNA species such that they chromatograph
markedly differently from the bulk of the tRNA [133].

Rechromatography

This is the most time consuming approach although it can yield many
tRNA species during the process. The crude tRNA is chromatographed on
BD-cellulose or RPC-5 etc., the desired fraction pooled and rerun on a
different column system or on the same column system using different solvent
conditions such as pH, absence of magnesium etc. For example pooling the
area of $tRNA_7^{Ser}$ acceptance of Fig. 10 [137] and rechromatographing it on an
RPC-5 column yields the profile shown in Fig. 11. This shows that the
first BD-cellulose column yielded a fraction containing mainly four tRNA
species. Rechromatography of this material produced a very pure fraction
of $tRNA_5^{Lys}$ and a relatively pure peak of $tRNA_7^{Ser}$ with a shoulder of $tRNA_6^{Ser}$
[139]. In this simple case two tRNA species are purified by two column
procedures and the tRNA is not exposed to any potential nuclease hazard or
chemical modification.

Purification by aminoacylation and derivitisation

The general procedure [133] involves the aminoacylation of the
desired tRNAs, chemical derivitisation such that the aminoacyl-tRNA becomes
hydrophobic and chromatography of the material on BD-cellulose which has
a high affinity for the derivitised aminoacyl-tRNA. The derivitisation
can be omitted if the amino acid being used is aromatic, although
derivatisation still yields a clearer separation.

The stages in the purification of $tRNA^{Phe}$ from Drosophila are shown
in Table 4 [140]. Crude tRNA was initially passed through a BD-cellulose
column to remove material that would normally only be eluted when ethanol
is added to the eluting buffer. The bulk of the tRNA eluting with 1.2 M
NaCl was charged with ^{14}C-phenylalanine and isolated by DEAE-cellulose
column chromatography as described under "Preparation of labeled aminoacyl-
tRNA". This was then derivitised with the 2 naphthoxyacetyl ester of
N-hydroxysuccinimide [133], and applied to a 20 x 3 cm BD-cellulose column

equilibrated with buffer C. This was successively eluted with the buffer
containing 1.0 M NaCl; 1.0 M NaCl + 5% ethanol; 1.0 M NaCl + 10% ethanol;
1.0 M NaCl + 20% ethanol and 1.0 M NaCl + 35% ethanol. The tRNA^{Phe} and
phenylalanyl-tRNA^{Phe} was eluted with the 1.0 M NaCl and 1.0 M NaCl + 5%
ethanol (non-derivatised ^{14}C-phenylalanyl-tRNA^{Phe} and uncharged tRNA^{Phe})
and the 1.0 M NaCl + 35% ethanol eluted the derivatised phenylalanyl-tRNA.

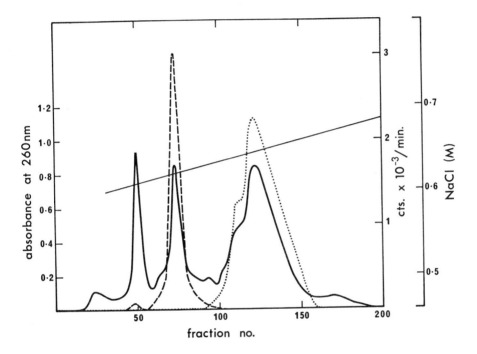

Fig. 11.   Re-chromatography of the pooled peak 7 fraction from Fig. 6, on
a RPC-5 column (60 cm x 2.4 cm). The column was eluted with a 2-1 gradient
of 0.6-0.7 M NaCl containing 0.01 mM MgCl$_2$, 0.01 M sodium acetate (pH 4.5)
and 1 mM 2-mercaptoethanol at 180 ml/h at 37°C; 10 ml fractions were
collected. ^{14}C-lysine ---- and ^{14}C-serine ········ acceptances were
determined by the paper disc procedure [139].

Naphthoxyacetylated aminoacyl-tRNAs usually elute at 1.0 M NaCl + 20%
ethanol but the presence of the aromatic amino acid produces an even
higher affinity for the BD-cellulose. The specific activity of this
material indicated it to be 50% pure. The tRNA^{Phe} was recovered from the
35% ethanol fraction by incubation in 1.8 M Tris-HCl buffer (pH 8.0) for
2 h at 22°C and chromatographed on a 100 x 0.9 cm BD-cellulose column
equilibrated with 0.35 M NaCl, 0.01 M MgCl$_2$ and eluted by a linear, 1-1
gradient from 0.3 to 1.2 M NaCl (Fig. 12).

Table 4   Stages in the Purification of Drosophila tRNA^{Phe}

Stage	A_{260} Units	% initial A_{260} units	Total pmoles a.a. accepted	% initial pmoles a.a. accepted	Spec.act. (pmoles/A_{260}) unit
Crude tRNA	18,600	100.0	543,340	100.0	31.9
BD-cellulose					
0.35 M NaCl	7,900	42.5	32,390	5.5	4.1
1.20 M NaCl	6,980	37.5	531,876	89.6	76.2
1.20 M NaCl-20% ethanol	1,900	10.2	17,670	3.0	9.3
Derivitisation of 1.20 M NaCl fraction					
1.0 M NaCl	4,706	25.3	133,423	22.5	28.4
1.0 M NaCl + 5% ethanol	986	5.3	39,592	6.7	40.4
1.0 M NaCl + 10% ethanol	186	1.0	905	0.2	5.0
1.0 M NaCl + 20% ethanol	707	3.8	2,272	0.4	3.2
1.0 M NaCl + 35% ethanol	242	1.3	229,945	38.8	942.4
BD-cellulose rechromatography of 1.0 M NaCl - 35% ethanol fraction after removal of the naphthoxyacetylphenylalanine					
Peak at 0.62 M NaCl	109	0.6	188,679	31.8	1731.0

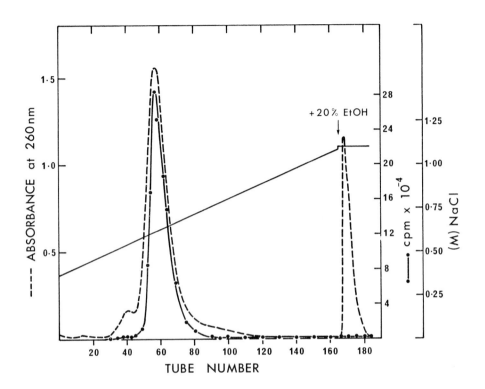

Fig. 12. Chromatography of <u>Drosophila</u> tRNA[Phe] on BD-cellulose. tRNA[Phe] (242 A_{260} units) obtained from the naphthoayacetylation procedure was applied to a 100 x 0.9 cm column and eluted by a (1-1) linear 0.35 - 1.2 $\underline{M}$ NaCl gradient followed by 20% ethanol containing 1.2 $\underline{M}$ NaCl. The flow rate was 50 ml/h with 5 ml fractions being collected. Acceptance of ^{14}C-phenylalanine was determined by the paper disc procedure [140].

Nuceloside analysis of tRNA species

      One of the interesting features of tRNAs is the content of modified bases and the relationship of the tRNAs that contain the same hyper-modified bases. One technique that requires very little material is the nucleoside tritium derivative approach [141, 142]. In this procedure the tRNA is enzymatically digested to nucleosides and the ribose oxidized with periodate. The aldehydes on the 2' and 3' carbons are then reduced and labeled with tritium using KB^3H_4. The labeled triols are then separated

by two-dimensional thin-layer chromatography and detected by autoradiography (Fig. 13). Identification of the modified nucleosides is based
solely on mobility during chromatography. An alternative approach [144]
that allows a better identification is the hydrolysis of at least 0.5 mg
of tRNA with RNase T_2 and separation of the nucelotides by two-dimensional
thin-layer chromatography. This provides sufficient material for a
spectral analysis of the modified bases.

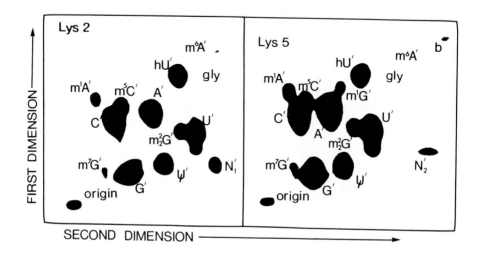

Fig. 13. Autoradiographic maps of ^{3}H-labeled nucleoside trialcohol
derivatives of <u>Drosophila</u> lysine tRNAs. The tRNAs were hydrolysed and the
nucleosides labeled and chromatographed by the method of Randerath <u>et al</u>.,
[141, 142]. Nucleosides are abbreviated according to the suggestions of
the IUPAC-IUB Commission on Biochemical Nomenclature (Progr. Nucleic Acid
Res. Mol. Biol. 9, IX, 1969.) The area marked b represents a background
spot and gly represents glycerol. N_1 is probably m^6A' and N_2 is most
likely derived from N_4 -acetylcytidine. The m^6A' is derived from the
breakdown of m'A'. [143].

    An alternative approach which gives an indication but not absolute
identification of certain modified nucleosides in tRNA species is to use
altered chromatographic properties of tRNAs following certain chemical
reactions. One of the most specific is the effect of $NaIO_4$ on tRNAs
containing the nucleoside Q [122, 145, 146]. Reaction of Q with periodate
causes formation of a very hydrophobic structure which increases the
affinity of the tRNA for both BD-cellulose and RPC-5 columns (Fig. 14)
resulting in retarded elution [145]. Reaction with cyanogen bromide can
give similar information indicating the presence of 2-thiouridine

derivatives [143, 147].

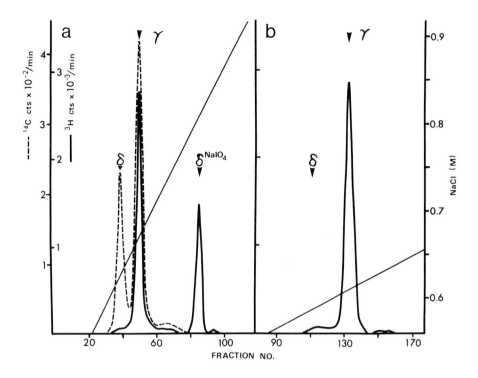

Fig. 14.   Effect of NaIO₄ on adult <u>Drosophila</u> ³H-tyrosyl-tRNA (a) Chromatography of NaIO₄ treated ³H-tyrosyl-tRNA and untreated ¹⁴C-tyrosyl-tRNA on an RPC-5 column eluted by a 0.55-1.50 M NaCl gradient.   (b) Chromatography of NaIO₄ treated ³H-tyrosyl-tRNA with a 0.55-0.70 M NaCl gradient [143].

δ indicates the tRNA containing the nucleoside Q and γ the tRNA lacking it.

Codon Recognition

The codon recognition of tRNAs is most easily achieved by the ribosome binding assay [148]. This can be performed without prior purification of the individual tRNA species [137]. Labeled aminoacyl-tRNAs are fractionated on RPC-5 columns and the individual isoaccepting peaks precipitated with ethanol, isolated on Millipore filters and eluted with 0.01 M potassium acetate pH 4.5. The relevant trinucleoside diphosphates can be conveniently synthesized enzymatically [149]. Such an analysis of <u>Drosophila</u> seryl-tRNA species is shown in Fig. 15.

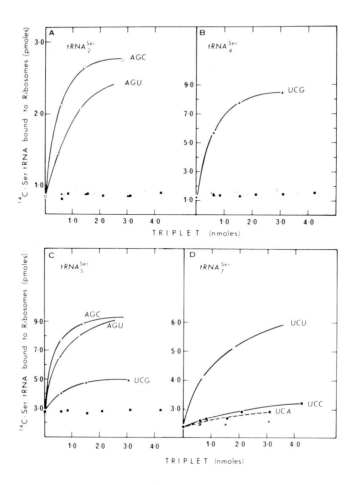

Fig. 15.   Binding of Drosophila ^{14}C-seryl-tRNAs to Eschericha coli ribosomes in the presence of triplets; UCC (■), UCU (△), UCA (●), UCG (▲), AGC (□), and AGU (○).   Reaction contained 0.02 M magnesium acetate, 0.05 KCl, and 0.1 M Tris-HCl, pH 7.2.   Reaction time was 15 min. Each 50- μl reaction contained (a) ^{14}C-seryl-tRNA$_2^{Ser}$ (14.4 pmol), (b) ^{14}C-seryl-tRNA$_5^{Ser}$ (19.4 pmol), and (d) ^{14}C-seryl-tRNA$_7^{Ser}$ (25.8 pmol). [137].

IX   CONCLUDING REMARKS

As insect biochemists pursue such problems as the mechanism of action of hormones and control of development more effort will be directed towards those molecular approaches that have been successfully applied to vertebrate systems.   We hope that this chapter will provide some guidance to those who decide RNA analysis and isolation will help solve their particular problems.

REFERENCES

1   Y. Suzuki and D.D. Brown, J. Mol. Biol., 63(1972)409.

2   P.M. Lizardi and D.D. Brown, Cell. 4(1975)207.

3   G.S. Chen and M.A.Q. Siddiqui, Proc. Nat. Acad. Sci., 70(1973)2610.

4   J.P. Garel, P. Mandel, G. Chavancy and J. Daillie, FEBS Lett., 7(1970)
    327.

5   J.P. Garel, J. Theor. Biol., 43(1974)211.

6   F.C. Kafatos, in E.W. Hanly (Editor), Problems in Biology: RNA in
    Development. University of Utah Press 1969, p.111.

7   R.E. Gelinas and F.C. Kafatos, Proc. Nat. Acad. Sci., 70(1973)3764.

8   J.N. Vournakis, R.E. Gelinas and F.C. Kafatos, Cell, 3(1974)265.

9   D.M. Glover, R.L. White, D.J. Finnegan and D.S. Hogness, Cell 5(1975)
    149.

10 S.W. Applebaum, R.P. Ebstein and G.R. Wyatt, J. Mol. Biol. 21(1966)29.

11 J. Shine and L. Dalgarno, J. Mol. Biol., 75(1973)57.

12 F.L. De Lucca, J.F. Giorgini and A. Calabrese, Experientia, 30(1974)149.

13 L. Dalgarno, D.M. Hosking and C.H. Shen, Eur. J. Biochem., 24(1972)498.

14 J.R.B. Hastings and K.S. Kirby, Biochem. J., 100(1966)532.

15 M. Adesnik and J.E. Darnell, J. Mol. Biol., 67(1972)397.

16 G. Brawerman, Ann. Rev. Biochem., 43(1974)621.

17 J.R. Greenberg, J. Cell. Biol., 64(1975)269.

18 P.M. Lizardi, R. Williamson and D.D. Brown, Cell, 4(1975)199.

19 E. Bard, D. Efron, A. Marcus and R.P. Perry, Cell, 1(1974)101.

20 R. Williamson, J. Crossley and S. Humphries, Biochemistry, 13(1974)703.

21 R.B. Church and F.W. Robertson, J. Exp. Zool., 162(1966)337.

22 R.E. McDowell, D. Feir and R.E. Ecker. Insect Biochem., 4(1974)295.

23 M.J. Burr and A.S. Hunter, B. Comp. Biochem. Physiol., 29(1969)647.

24 A. Fleck and H.N. Munro, Biochim. Biophys. Acta, 55(1962)571.

25 B. Linzen and G.R. Wyatt, Biochim. Biophys. Acta, 87(1964)188.

26 Y. Chinzei and S. Tojo, J. Insect. Physiol., 18(1972)1683.

27 Z. Dische, in E. Chargaff and J.N. Davidson (Editors) The Nucleic Acids
    Vol. 1, Academic Press, New York 1955, p. 300.

28 S.E. Kerr and K. Seraidarian, J. Biol. Chem., 159(1945)211.

29 G. Ceriotti, J. Biol. Chem., 214(1955)59.

30 W.C. Schneider, in S.P. Colowick and N.O. Kaplan, (Editors), Methods
    in Enzymol. Vol. 3(1957)680.

31 O.H. Lowry, N.R. Roberts, K.Y. Leiner, M.L. Wu and A.L. Farr, J. Biol.
    Chem., 207(1954)1.

32   G. Schmidt, in S.P. Colowick and N.O. Kaplan (Editors) Methods in
     Enzymol., Vol. 3, Academic Press, New York, 1957, p.671.

33   C.P. Emerson and T. Humphreys, Anal. Biochem., 40(1971)254.

34   A. Fausto-Sterling, L.M. Zheutlin and P.R. Brown, Devel. Biol. 40
     (1974)78.

35   H.H. Hagedorn, A.M. Fallon and H. Laufer, Devel. Biol. 31(1973)285.

36   A. Spradling, R.H. Singer, J. Lengyel and S. Penman, in D.M. Prescott
     (Editor) Methods in Cell Biol. Vol. 10, Academic Press, New York,1975.

37   J. Lengyel, A. Spradling and S. Penman, in D.M. Prescott (Editor),
     Methods in Cell Biol. Vol. 10, Academic Press, New York, 1975.

38   J. Lengyel and S. Penman, Cell, 5(1975)281.

39   D.E. Wimber and D.M. Steffensen, Science, 170(1970)639.

40   M.S. Kaulenas, J. Insect Physiol., 16(1970)813.

41   M.S. Kaulenas, J. Insect Physiol., 18(1972)649.

42   H. Fraenkel-Conrat, B. Singer and A. Tsugita, Virology, 14(1961)54.

43   B. Singer and H. Fraenkel-Conrat, Virology, 14(1961)59.

44   C.G. Rosen and I. Fedorcsak, Biochim. Biophys. Acta, 130(1966)401.

45   N.J. Leonard, J.J. McDonald and M.E. Reichmann, Proc. Nat. Acad. Sci.,
     67(1970)93.

46   R.D. Palmiter, Biochemistry, 13(1974)3606.

47   M. Denic, L. Ehrenberg, I. Fedorcsak and F. Solymosy, Acta Chem.
     Scand. 24(1970)3753.

48   R.T. Schimke, R. Palacios, D. Sullivan, M.L. Kiely, C. Gonzales and
     J.M. Taylor, in L. Grossman and K. Moldave (Editors) Methods in
     Enzymol. Vol. 30, Part F, Academic Press, New York, 1974, p. 631.

49   A.J. Howells and G.R. Wyatt, Biochim. Biophys. Acta, 174(1969)86.

50   J.S. Roth, Biochim. Biophys. Acta, 21(1956)34.

51   K. Shortman, Biochim. Biophys. Acta, 51(1961)37.

52   G. Blobel and V.R. Porter, J. Mol. Biol., 28(1967)539.

53   G.R. Wyatt - "personal communication".

54   C.L. Hew and C.C. Yip, Can. J. Biochem., 52(1974)959.

55   U. Wiegers and H. Hilz, Biochem. Biophys. Res. Comm. 44(1971)513.

56   U.Z. Littauer and M. Sela, Biochim. Biophys. Acta, 61(1962)609.

57   L. Marcus and H.O. Halvorson, In L. Grossman and K. Moldave (Editors)
     Methods in Enzymol. Vol. 12, Part A, Academic Press, New York, 1968,
     p. 498.

58   L.A. Grivell, L. Reijnders and P. Borst, Biochim. Biophys. Acta,
     247(1971)91.

59 E.R.M. Kay, N.S. Simmons and A.L. Dounce, J. Amer. Chem. Sco., 74
    (1952)1724.

60 H. Noll and E. Stutz, in L. Grossman and K. Moldave (Editors), Methods
    in Enzymol., Vol. 12, Part B, Academic Press, New York, 1968, p. 129.

61 K.S. Kirby, Biochem. J., 64(1956)405.

62 K.S. Kirby, in L. Grossman and K. Moldave (Editors), Methods in
    Enzymol., Vol. 12, Part B, Academic Press, New York, 1968, p.87.

63 K.S. Kirby, Biochem. J., 96(1965)266.

64 K. Scherrer, in K. Habel and N.P. Slazman (Editors), Fundamental
    techniques in virology, Academic Press, New York, 1969, Ch.38, p.413.

65 A. Gierer and G. Schramm, Nature, 117(1956)702.

66 A. Sibatani, S.R. de Kloet, V.G. Allfrey and A.E. Mirsky, Proc. Nat.
    Acad. Sci. 48(1962)471.

67 G.P. Georgiev and V.L. Mantieva, Biochim. Biophys. Acta, 61(1962)153.

68 K.S. Kirby, Biochim. Biophys. Acta, 55(1962)545.

69 M.G. Sevag, D.B. Lackmann and J. Smolens, J. Biol. Chem., 124(1938)425.

70 F.L. De Lucca and M.T. Imaizumi, Biochem. J., 130(1972)335.

71 J.F. Giorgini and F.L. De Lucca, Biochem., J., 135(1973)73.

72 F.L. De Lucca, M.T. Imaizumi and A. Haddad, Biochem. J., 139(1974)151.

73 R.A. Cox, in L. Grossman and K. Moldave (Editors), Methods in Enzymol.
    Vol. 12, Part B, Academic Press, New York, 1968, p. 120.

74 M. Sela, C.B. Anfinsen and W.F. Harrington, Biochim. Biophys. Acta, 26
    (1957)502.

75 J. Barlow, and A.P. Mathia, in G.L. Cantoni and D.R. Davies (Editors),
    Procedures in nucleic acid research, Harper and Row, New York, 1966,
    p. 444.

76 J.J. Barlow, A.P. Mathias, R. Williamson and D.B. Cammack, Biochem.
    Biophys. Res. Commun., 13(1963)61.

77 H. Ishikawa and R.W. Newburgh, J. Mol. Biol., 64(1972)135.

78 J.R. Greenberg, J. Mol. Biol., 46(1969)85.

79 J.R.B. Hastings and K.S. Kirby, Biochem. J., 100(1966)532.

80 R.D. Palmiter, J. Biol. Chem., 248(1973)2095.

81 D.D. Brown and Y. Suzuki, in L. Grossman and K. Moldave (Editors)
    Methods in Enzymol., Vol. 30, Part F, Academic Press, New York, 1974,
    p. 648.

82 K. Moldave and L. Skogerson, in L. Grossman and K. Moldave (Editors),
    Methods in Enzymol., Vol. 12, Part A, Academic Press, New York, 1967,
    p. 478.

83 E.H. McConkey, In L. Grossman and K. Moldave (Editors), Methods in Enzymol., Vol. 12, Part A., Academic Press, New York, 1967, p. 620.

84 C.W. Dingman and A.R. Peacock, Biochemistry, 7(1968)659.

85 F.L. De Lucca and G.R. Wyatt, "unpublished results".

86 H. Aviv and P. Leder, Proc. Nat. Acad. Sci., 69(1972)1408.

87 M. Hirsch, A. Spradling and S. Penman, Cell, 1(1974)31.

88 A. Spradling, H. Hui and S. Penman, Cell, 4(1975)131.

89 U. Lindberg and T. Persson, Eur. J. Biochem., 31(1972)246.

90 M. Adesnik, M. Salditt, W. Thomas and J.E. Darnell, J. Mol. Biol., 71 (1972)21.

91 U. Lindberg, T. Persson and L. Philipson, J. Virol., 10(1972)909.

92 S.Y. Lee, J. Mendecki and G. Brawerman, Proc. Nat. Acad. Sci. 68 (1971)1331.

93 P.M. Lizardi, Cell, 7(1976)239.

94 T.T. Chen, P. Couble, F.L. De Lucca, and G.R. Wyatt, in L.I. Gilbert (Editor), The Juvenile Hormones, "in press".

95 G. Brawerman, J. Mendecki and S.Y. Lee, Biochemistry, 11(1972)637.

96 K. Scherrer and J.E. Darnell, Biochem. Biophys. Res. Commun., 7(1962) 486.

97 S. Penman, J. Mol. Biol., 17(1966)117.

98 R.P. Perry, J. La Torre, D.E. Kelley and J.R. Greenberg, Biochim. Biophys. Acta, 262(1972)220.

99 A. Spradling, S. Penman and M.L. Pardue, Cell, 4(1975)395.

100 R. Palacios, R.D. Palmiter and R.T. Schimke, J. Biol. Chem., 247 (1972)2316.

101 P.V. Vignais, B.J. Stevens, J. Huet and J. Andre, J. Cell. Biol., 54 (1972)468.

102 P. Couble, F. L. De Lucca and G.R. Wyatt, "unpublished results".

103 A. Efstratiadis, T. Maniatis, F.C. Kafatos, A. Jeffrey and J.N. Vournakis, Cell, 4(1975)367.

104 A. Efstratiadis and F.C. Kafatos in J. Last (Editor) Methods in Molecular Biology, Vol. 8, Marcel Dekker, New York, "in press".

105 A. Efstratiadis, F.C. Kafatos, A.M. Maxam and T. Maniatis, Cell, 7 (1976)279.

106 H. Lerer and B.N. White, "unpublished results".

107 G.C. Rosenfeld, J.P. Comstock, A.R. Means and B.W. O'Malley, Biochem. Biophys. Res. Commun., 46(1972)1695.

108 R.E. Rhoads, G.S. McKnight and R.T. Schimke, J. Biol., Chem., 248(1973) 2031.

109 I. Schechter, Proc. Nat. Acad. Sci., 70(1973)2256.

110 L. Eron and H. Westphal, Proc. Nat. Acad. Sci., 71(1974)3385.

111 B.E. Roberts and B.M. Paterson, Proc. Nat. Acad. Sci., 70(1973)2330.

112 J.W. Davies and P. Kaesberg, J. Virol., 12(1973)1434.

113 I. Gozes, H. Schmitt and U.Z. Littauer, Proc. Nat. Acad. Sci., 72
     (1975)701.

114 R.A. Greene, M. Morgan, A.J. Shatkin and L.P. Gage, J. Biol. Chem.,
     250(1975)5114.

115 E.G. Fragoulis and C.E. Sekeris, Eur. J. Biochem., 51(1975)305.

116 J.B. Gurdon, C.D. Lane, H.R. Woodland, and G. Marbrix, Nature 233
     (1971)177.

117 T.T. Chen, H. Lerer, B.N. White, G.R. Wyatt - "unpublished results".

118 U.K. Laemmli, Nature, 227(1970)680.

119 D.R. Senger and P.R. Gross, Devel. Biol. (1976) "in press".

120 B.N. White, G.M. Tener, J. Holden and D.T. Suzuki, Devel. Biol., 33
     (1973)185.

121 B.N. White, G.M. Tener, J. Holden and D.T. Suzuki, J. Mol. Biol., 74
     (1973)635.

122 B.N. White, Biochim. Biophys. Acta, 353(1974)283.

123 N.J. Lassam, H. Lerer and B.N. White, Nature, 256(1975)734.

124 A.D. Kelmers, G.D. Novelli and M.P. Stulberg, J. Biol. Chem., 240
     (1965)3979.

125 K.H. Muench and P. Berg in G.L. Cantoni and D.R. Davis (Editors),
     Procedures in Nucleic Acid Research, Harper and Row, New York, 1966,
     p. 375.

126 B.N. White and G.M. Tener, Can. J. Biochem. 51(1973)896.

127 N.J. Lassam, H. Lerer, B.N. White, Devel. Biol. 49(1976)268.

128 F.G. Bollum, J. Biol. Chem. 234(1959)2733.

129 B.N. White, N.J. Lassam and H. Lerer, in L.I. Gilbert (Editor), The
     Juvenile Hormones, "in press".

130 M. Wosnick and B.N. White - "unpublished results".

131 W.K. Yang and G.D. Novelli, Biochem. Biophys. Res. Commun., 31(1968)
     534.

132 R.L. Pearson, J.F. Weiss and A.D. Kelmers, Biochim. Biophys. Acta,
     228(1971)770.

133 I. Gillam, S. Millward, D. Blew, M. von Tigerstrom, E. Wimmer and G.M.
     Tener, Biochemistry, 6(1967)3043.

134 B.N. White and G.M. Tener, Anal. Biochem., 55(1973)394.

135 J.Ilan, J.Ilan and N. Patel, J. Biol. Chem, 245(1970)1275.

136   J. Ilan and J. Ilan, Devel. Biol. 42(1975)64.

137   B.N. White, R. Dunn, I. Gillam, G.M. Tener, D.J. Armstrong, F. Skoog,
      C.R. Frihart, and N.J. Leonard, J. Biol. Chem., 250(1975)515.

138   J.D. Cherayil, A. Hampel and R.M. Bock, in S.P. Colowick and N.O.
      Kaplan (Editors), Methods in Enzymol. Vol. 12, Part B, 1968, p. 166.

139   T.A. Grigliatti, B.N. White, G.M. Tener, T.C. Kaufman, and D.T. Suzuki,
      Proc. Nat. Acad. Sci., 71(1974)3527.

140   B.N. White and G.M. Tener, Biochim. Biophys. Acta, 312(1973)267.

141   E. Randerath, C.T. Yu and K. Randerath, Anal. Biochem. 48(1972)172.

142   K. Randerath, E. Randerath, L.S.Y. Chia, and B.J. Nowak, Anal.
      Biochem., 59(1974)263.

143   B.N. White, Biochim. Biophys. Acta, 395(1975)322.

144   S. Nishimura in N.J. Davidson and W.E. Cohn (Editors) Progress in
      Nucleic Acid Research and Molecular Biology, Vol. 12, Academic Press,
      New York, 1972, p.49.

145   F. Harada and S. Nishimura, Biochemistry, 11(1972)301.

146   H. Kasai, Z. Ohashi, F. Harada, S. Nishimura, N.J. Oppenheiner, P.F.
      Crain, J.G. Liehr, D.L. von Minden and J.A. McCloskey, J. Biol. Chem.,
      14(1975)4198.

147 ·M. Saneyoshi and S. Nishimura, Biochim. Biophys. Acta, 204(1970)389.

148   M. Nirenberg and P. Leder, Science 145(1964)1399.

149   R.E. Thach and P. Doty, Science, 148(1965)632.

CHAPTER 4

ANALYSIS OF AMINO ACIDS, PEPTIDES AND RELATED COMPOUNDS

P. S. Chen

Institute of Zoology, University of Zürich, Switzerland

CONTENTS

INTRODUCTION

Insects are known to be characterized by an unusually high
concentration of free amino acids in both tissue and hemolymph.
Compared with human blood, the total content of these compounds
in insect hemolymph is from 50 to 300 times higher (1, 2).
They account for about 50 to 85% of the total non-protein
nitrogen in the blood of various insects(3). As discussed in
previous reviews (4,5), one possible explanation of such a
high titre is that the amino acids in insects may fulfill a
number of specific functions aside from serving as substrates
for protein synthesis. These include regulation of the osmotic
pressure, detoxication of waste products, acting as energy
sources under special conditions and participation in a variety
of morphogenetic and reproductive processes.

Before the advent of paper partition chromatography qualitative and quantitative studies of amino acids and related
compounds in insects were limited, obviously owing to the lack
of suitable methods. Since then a large volume of information
has become available (6). As in other animals, all amino acids
which commonly occur in proteins are also present in insects.
In addition, several amino acids such as ornithine, β-alanine,
taurine, methionine sulphoxide, α-, β- and γ-aminobutyric acid,
which do not enter into protein molecules, have been identi-
fied. Of further interest is the occurrence of D-isomers such
as D-alanine in the milkweed bug Oncopeltus fasciatus (7) and
the monarch butterfly Danaus plexippus (8) and D-serine in the
silkworm Bombyx mori (9,10). Other amino acids which have been
found only in a single species are S-methylcysteine (11), homo-
arginine (12) and methylhistidine (13,14).

In recent years simple and sensitive methods with high
resolving power have been rapidly developed. The present chapter is a description of some of those methods which have been
extensively used in the author's laboratory for studies of
amino acids in insects. Although our experience comes largely
from work on dipterans, with slight modifications these methods
can be easily applied to other insect species. No attempt has
been made to provide a general treatment of all available techniques but rather the tried and proven procedures are summarized
as a guide for those who use insects as working material. There
exist many excellent monographs and manuals devoted specifically to chromatographic and electrophoretic techniques (15-23)
which can be consulted for more detailed information and
further references.

PREPARATION OF EXTRACTS

Extraction of amino acids for paper chromatography and
electrophoresis

Since insects have a hard cuticle, it is of advantage to
use an all-glass homogenizer to extract the ninhydrin-reacting components. Grinding the pupae or adult flies in a conventional homogenizer with a Teflon pistil yields usually
large pieces of chitin with attached tissues, resulting in
incomplete extraction. Small samples for use in paper-chroma-
tographic or electrophoretic analysis can be conveniently
prepared by employing a conical microhomogenizer, shown in
Fig. 1. With some practice it can be readily made in the laboratory from a Sorvall centrifuge tube of 1 ml volume. To do
this, a trace of silicon carbide is added to the tube and then
ground gently with a glass rod which has previously been
drawn on a small flame to the proper shape.

A few pieces of tissues or organs dissected from either
larvae or flies are added to about 20 μl pre-cooled 80% methanol contained in the microhomogenizer which is in turn held in
an ice bath. The collected material is then homogenized for a
few minutes with either the hand or preferably with a motor
driven at median speed. Following the addition of 100 μl 80%

methanol, the homogenate is centrifuged at $0°C$ for 10-15 min at
ca. 6'000 x g. The supernatant fraction is carefully pipetted
out, the residue washed with another 100 μl 80% methanol and
again centrifuged. The pooled methanolic extract can be used for
chromatographic or electrophoretic separation on filter paper
without further purification.

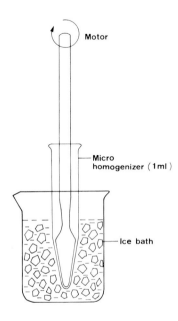

Fig. 1. All-glass microhomogenizer for preparing extracts
from insect tissues or organs.

    The extraction procedure described above is also adequate
for preparing samples from 10-15 mg of whole larvae, pupae or
adults. With the same type of homogenizer of a larger volume
(5-10 ml) 3-5 g of live material can be extracted in a similar
manner.
    Extraction of amino acids for ion-exchange chromatography
    The method for preparing extracts from about 10-50 g of
live or frozen material is illustrated in Fig. 2. It is a
simplified diagram of the procedure previously described by
Mitchell and Simmons (24) for differential extraction of amino

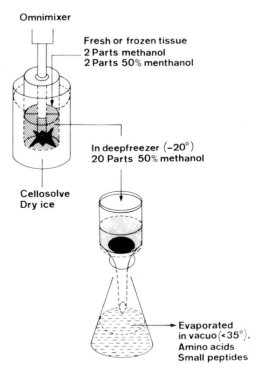

Fig. 2. Diagram of procedures for extracting amino acids, small peptides and related compounds from insects. (Redrawn and simplified from Mitchell and Simmons(24).)

acids, lipid derivatives, large peptides and proteins from Drosophila larvae. The animals collected are transferred to the container of a Servall Omnimixer or Bühler homogenizer, 2 parts of pre-cooled methanol (100%) are added to one part of material, and the suspension is homogenized at maximum speed for 10-15 min. During homogenization the container is kept in a dry ice-cellosolve bath. After the addition of another 2 parts of 50% methanol the resulting slurry is filtered on sintered glass at -20°C, and the residue washed with a total of 20 parts of cold 50% methanol. The filtrate is then evaporated in a flash evaporator at 26-30°C, and the dry residue dissolved in distilled water to a final concentration of approximately 2g/ml. To remove the lipid the resulting solution is transferred into a separatory funnel and shaken twice with one part of chloroform.

Following separation, the water phase is retained, centrifuged
again if necessary, and the clear extract stored in the cold.

Amino acid sample from hemolymph

For analysis of amino acids in the larval hemolymph, the
animals are washed thoroughly with water, dried on a piece of
absorbant paper and opened carefully with two dissecting needles
on a cleaned slide under the binocular microscope. The hemolymph that flows out is immediately aspirated into a micropi-
pette and collected in a pre-cooled glass tube containing a
few crystals of phenylthiourea. The latter is a potent inhibitor
of the enzyme tyrosinase. Following removal of the hemocytes by
centrifugation, the collecting tube is heated for 1-2 min in a
water bath (100°C). The sample is then briefly stirred with a
fine glass rod and again centrifuged. The supernatant fraction
is retained for amino acid analysis. If necessary, the residue
can be washed once with a small volume of water and the supernatant pooled. Our experience shows that protein denaturation
by brief heating causes no detectable loss of ninhydrin-positive material.

Comments

a. Temperature - As in other biological materials, in the
preparation of amino acid extracts from insects it is essential
that peptides, phosphate esters and lipid conjugates do not
undergo significant hydrolysis during the extraction procedure.
It is found that incubation of homogenate from Drosophila larvae at 28°C leads to rapid increase in ninhydrin-reacting components within a few minutes (Fig. 3). Thus it is advisable to
carry out the extraction at low temperature in the range of
0° to -20°C.

b. Deproteinization - For a satisfactory separation of amino
acids and related compounds by either of the methods described
below, the sample should be essentially free of protein. In
the extraction procedure mentioned above this is achieved by
methanol at a final concentration of 80%. Even at this concentration deproteinization may not be complete. Further denaturation can be effected by dipping the homogenizer briefly

into boiling water, though this is usually not necessary. The
use of methanol or ethanol to prepare the extract has the advantages that it can be evaporated easily and causes no decomposition of the components to be analyzed. For chromatography on
filter paper the alcoholic sample can be directly spotted.

Other widely used methods of deproteinization include the
addition of an appropriate volume of 10% trichloracetic acid,
10% perchloric acid or 1% picric acid (17). Trichloracetic acid
in the extract can be removed by extraction with ether, and perchloric acid by the addition of 50% potassium hydroxide followed
by removing the resulting insoluble potassium perchlorate by
centrifugation. The removal of picric acid can be accomplished
by passing the sample through a small column of Dowex 2-x10 resin.

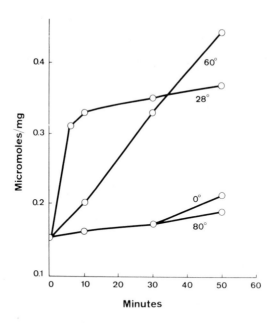

Fig. 3. Increase of ninhydrin-positive components by
incubating homogenates of 2nd instar Drosophila larvae at
different temperatures. (From Mitchell and Simmons (24).)

It is clear that all these treatments are more tedious and
time-consuming than the use of methanol or ethanol, and need
be performed under more carefully controlled conditions.

  c. Pigments, lipids, carbohydrates and salts - Insects
contain various soluble colored components, mostly ommochromes.
But they do not show serious interference in either the chromatographic separation or the ninhydrin reaction. Ommochrome
adsorbed on the resin is usually removed during regeneration
of the column. Removal of the pigments by treating the extract
with active charcoal causes a significant loss of various ninhydrin-reacting materials in particular the aromatic amino acids
such as tyrosine and phenylalanine.

  Fat and fatty acids do not interfere in the preparation of
the sample, as most of these compounds are insoluble in aqueous
solution and those that are soluble can be easily removed by
chloroform. In our experience for the use of small samples on
filter paper removal of lipids prior to analysis is not necessary. On the other hand, for column chromatography many of
these compounds bind irreversibly to the resin and can only be
removed by rather elaborate procedures.

  Carbohydrates also do not interfere in either the preparation
of the extract or the subsequent analysis. The only difficulty
is that, if the sample is subjected to acid hydrolysis, large
quantities of colored precipitates are formed and have to be
removed by centrifugation before analysis.

  Appreciable quantities of dissolved salts in the extract result in an incomplete separation of the components by paper
chromatography, and should be kept at a minimum in the preparation of the sample. However, a higher salt concentration can
be tolerated in ion-exchange chromatography. In general, samples
containing about 5-10 mM of dissolved salts cause no significant distortion of the peaks in the elution profile. But ammonia,
free or in the form of ammonium salts, imposes difficult problems if the sample is chromatographed by the ion-exchange
procedure. When sufficient material is available, removal of
all ammonia prior to analysis is always recommended.

TWO-DIMENSIONAL PAPER CHROMATOGRAPHY

One obvious reason that paper chromatography has been most widely used for the study of insect amino acids is due to its high sensitivity. Owing to the small size of most insects, the available material is limited, especially when the analysis of individual tissues or organs is required. As summarized in Table 1, results of reasonable accuracy can be obtained with quantities of amino acids as little as 0.05 µg (25).

Under normal conditions the experimental procedure is simple. The Rf-values of virtually all known amino acids and derivatives are available and may serve as a guide to the choice of the chromatographic system. It is, however, not always possible to predict on the basis of these data which system is most suitable of the analysis of the sample in question. This is especially the case for insects which show the largest quali- tative and quantitative diversities of amino acids and related compounds. Thus it is advisable to try several separation pro- cedures in preliminary experiments prior to undertaking a long- term chromatographic study.

We found that 10 to 20 µl of hemolymph or tissue extracts equivalent to 10-15 mg of live material are optimal for a two- dimensional separation. The sample is spotted on a sheet of Whatman No. 1 filter paper (24 x 46 cm) with a stream of warm air to keep the starting point as small as possible. Two sepa- ration systems have been shown to yield the most satisfactory results. In the first system the paper is run in 70% n-propanol in the ascending direction for 10-14 h at room temperature (20- 22°C), and followed by a second run in water-saturated phenol in the descending direction for 20-22 h. In the second system phenol saturated with 0.1 M borate buffer (pH 8,6) is used as the first solvent (descending) and n-butanol:acetic acid:water in a mixture of 4:1:5 as the second solvent (ascending). Following the phenol run, the paper is dried for 2½ h at 60°C. A longer heating of the wet chromatogram at this temperature leads to an increasing loss of the amino acids. Subsequent to the propanol or butanol-aceticacid-water run, the paper is dri- ed for 15-30 min in the hood in a current of air at room

Table 1
Sensitivity of ninhydrin reaction for detection of
amino acids on two-dimensional paper chromatogram
(From Auclair and Dubreuil (25).)

Amino acid	Minimum quantity detected µg
Alanine	0.06
α-Amino-n-butyric acid	0.12
Arginine	4.0
Asparagine	0.8
Aspartic acid	0.2
β-Alanine	0.22
Cysteic acid	0.2
Glutamic acid	0.1
Glutamine	0.4
Glycine	0.05
Histidine	7.5
Hydroxyproline	1.0
Leucine and isoleucine, 1:1	0.25
Lysine	1.5
Methionine sulphoxide	0.5
Phenylalanine	1.25
Proline	1.5
Serine	0.08
Taurine	0.2
Threonine	0.2
Tryptophan	2.0
Tyrosine	1.0
Valine	0.15

temperature. As insect contain àn abundance of fluorescent sub-
stances(26), a preliminary evaluation of the separation procedure can be made by viewing the chromatogram under ultraviolet
light after the propanol run.

The dried chromatogram is pulled quickly through a solution
of 0.5% ninhydrine in acetone and left standing for 20 min in
the air. The color reaction is developed at 60°C for 30 min.
Various staining reagents containing 0.1 to 0.5% ninhydrin dissolved in propanol, ethanol, ether or water-saturated n-butanol
with or without the addition of acetic acid have been used (17).
Our experience shows that these modifications yield essentially
the same results. Development of the ninhydrin color by heating
the chromatogram at 80 to 105°C for various lengths of time
have also been suggested. But at such high temperatures many
amino acids are partially destroyed. This may account for the
wide range of sensitivity found for individual amino acids.

The ninhydrin reaction suffers from the disadvantage that
the colored spots on the chromatogram are unstable. For quantitative estimation, the developed color is converted to the more
stable Cu- or Cd-complex. A convenient method is that given by
Fischer and Dörfel (27). The procedures are as follows: The
colored spots are cut out of the chromatogram, placed individually in a test tube and extracted for at least 2 h with 5 ml
of a copper nitrate solution (2 ml of water-saturated $Cu(NO_3)_2$
plus 0.2 ml of 10% HNO_3 diluted to 500 ml with absolute methanol). The color intensity of the extract is read at 510 nm.

The patterns of ninhydrin-reacting components in male flies
of Drosophila melanogaster obtained by using the two chromatographic systems described above are illustrated in Fig. 4. A
male-specific substance previously located in the paragonial
gland (28) shows overlap with glutamine in the propanol-phenol
system, but is distinctly separated from the latter by n-buta-
nol:acetic acid:water. Various phosphate esters such as phosphoserine, phosphoethanolamine and tyrosine phosphate, which
are frequently encountered in insect material, have in general
low Rf-values in both solvents. Small peptides, if present,

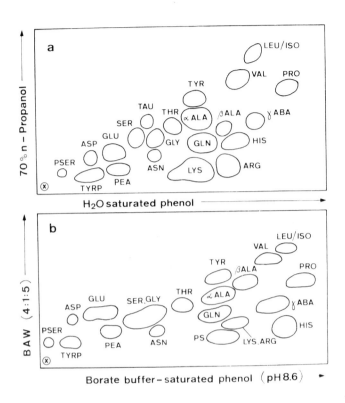

Fig. 4. Two-dimensional paper chromatograms of amino acids
and derivatives in male flies of D. melanogaster by using
different solvents. For abbreviations, see legend to Fig. 10.

occur usually in the region between aspartic acid, glutamic
acid and serine.

FINGERPRINT ANALYSIS

The use of phenol as solvent has two disadvantages: it de-
stroys a considerable amount of the amino acids and has a slow
flowing rate. Both handicaps can be eliminated by fingerprint
analysis, i.e. high-voltage electrophoresis in the first dimension, followed by paper chromatography in the second dimension.
Excellent separation can be accomplished by the following
procedure:

The sample prepared from about 15-20 mg material is applied

to a sheet of Whatman 3MM filter paper (32 x 57 cm) which is
then sprayed thoroughly with 8% formic acid (pH 1.5). Following
this, the excess formic acid is removed by pressing the paper
on a clean dry sheet. The paper is carefully placed on the glass
plate in an electrophoresis apparatus (Pherograph 64, L. Hor-
muth, Germany), and subjected to separation with a current of
1800 to 2000 V and 100 to 120 mA for approximately 2 h. During
the electrophoretic run the cooling chamber is maintained at
-6°C. Subsequent to evaporation of the formic acid by heating
the paper at 65°C for $2\frac{1}{2}$ to 3 h, separation in the second dimension is performed by ascending chromatography with either
70% n-propanol or n-butanol:acetic acid:water (4:1:5) for ca.
22 h at $22-23^{\circ}$C. Development of the ninhydrin color and quantitation of the amino acids are carried out in the same way as
that given for the two-dimensional paper chromatography.

The method described above is simple and has a high resolving power. The only precaution is that, because of the heat
development on the paper during the electrophoretic run, an
efficient cooling system is required. A higher sensitivity
can be achieved by using the thiner Whatman No. 1 paper, which,
however, has to be handled with special care after being sprayed with formic acid.

By employing the fingerprint procedure we have successfully
separated various ninhydrin-positive compounds in the male
adults of D. funebris and D. melanogaster as well as one not
yet identified tyrosine derivative in the larva of D. busckii
(Fig. 5-7). The essential step of the separation is based on
the distinctly different electrophoretic mobilities of these
compounds. All acidic components (phosphoserine, phosphoethanolamine and glycerophosphoethanolamine) remain near the start
point, whereas the amino acids move rapidly in the cathodal
direction.

TWO-DIMENSIONAL HIGH-VOLTAGE PAPER ELECTROPHORESIS

Peptides contribute to a considerable part of free ninhydrin-
positive components in insects. Both Drosophila and Culex have

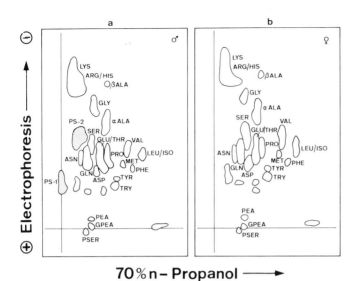

Fig. 5 Fingerprint analysis of amino acids and derivatives
in methanolic extracts from 30 male (a) and 20 female abdomens
of D. funebris. (From H. Baumann und Chen (63).)

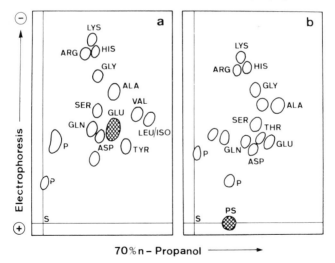

Fig. 6. Fingerprint analysis of amino acids and derivatives
in methanolic extracts from 200 paragonial glands of adult
males of D. nigromelanica (a) and 160 paragonial glands of
adult males of D. melanogaster (b). (From Chen and Oechslin
(64).)

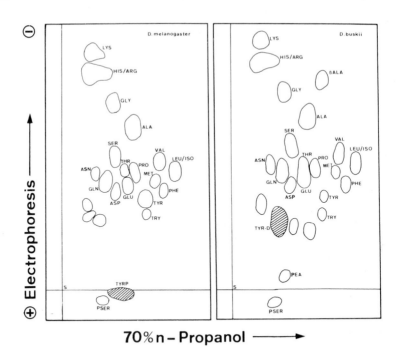

Fig. 7. Fingerprint analysis of amino acids and derivatives in methanolic extracts from 20 mature larvae of D. melanogaster (left) and D. busckii (right). (From Chen (65).)

been shown to posses many acidic peptides (24, 30), while in the larval hemolymph of the blowfly Phormia regina the occurrence of various basic peptides has been demonstrated (31). The similar composition, large number and low concentration of most of these compounds impose difficult problems on the separation and identification. As noted previously, on paper chromatograms all acidic peptides in Drosophila move very slowly in the solvents used and accumulate in positions occupied by aspartic acid and glutamic acid (29). Similarly the basic peptides in Phormia overlap to a large extent with lysine, arginine and histidine. According to our experience more satisfactory separation can be accomplished by two-dimensional paper electrophoresis.

In general, 20 μl hemolymphe or a tissue extract equivalent to about 30 mg wet weight is applied to a sheet of Whatman 3 MM filter paper (37 x 41 cm). The method of electrophoretic sepa-

ration is similar to that used by Gross (32). Following an initial separation in 8% formic acid at pH 1.5 and 1300-1400 V
for 1½ h, the paper is thoroughly dried and subjected to another run in the second dimension in a carbonate-bicarbonate
buffer (17 g Na_2CO_3 + 28.6 g Na HCO_3 + 10 l H_2O) at pH 8.6 and
2000-2100 V for 1¼ h. All conditions for treating the paper
with the buffer solution, maintainance of the cooling system,
development of the ninhydrin color and quantitative estimation
of the compounds on the electropherogram are the same as those
described in the previous section.

One useful method for detecting the peptides separated on
the paper by electrophoresis is the hypoclorite-starch-iodide
test of Rydon and Smith (33) as modified by Pan and Dutcher (34).
The electropherogram is first stained with the ninhydrin reagent
as usual, and all colored spots, known or unknown, are encir-
cled with a pencil. The developed paper is then sprayed with
a 0.28% sodium hypochlorite solution, following which the ninhydrin color disappears. The paper is dried in the air and again
sprayed with 96% ethanol to remove the excess hypochlorite. It
is of advantage to repeat this procedure twice in order to remove the hypochlorite completely. After evaporation of the ethanol, the paper is subjected to a final spray with a mixture
containing equal volumes of a 1% starch and a 1% potassium
iodide solution. All peptide spots immediately give a deep blue
color according to the following reaction:

$$-CO.NH- + Cl_2 \longrightarrow -CO.NCl- \;+\; KI \longrightarrow -CO.NH- \;+\; KCl \;+\; \tfrac{1}{2}\,I_2$$

The peptide pattern of the hemolymph from the early 3rd instar larvae of Phormia is depicted in Fig. 8 and that of a
tissue extract from the mutant l(3)tr of Drosophila in Fig. 9.
In both cases the electropherogram was first treated with ninhydrin (upper photograph) and then sprayed with the hypochlorite-
starch-iodide reagent (lower photograph). The results confirmed the previous findings that Phormia larvae contain predominantly basic peptides, whereas all peptides in Drosphila are
acidic (35). It is worthy of note that in the Drosophila extract

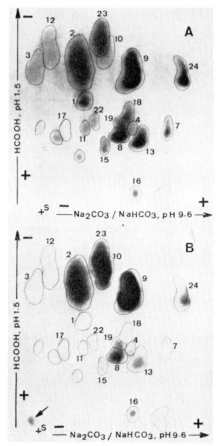

Fig. 8. Separation of amino acids, peptides and related compounds in the larval hemolymph of Phormia regina by two-dimensional paper electrophoresis. The electropherogram was first
stained with 0.5% ninhydrin (A) and then sprayed with hypochlorite and starch-iodide reagents (B). The spots are numbered
as follows: 1,α-alanine; 2, β-alanine + peptide; 3, arginine;
4, asparagine; 5, aspartic acid; 6, cysteine; 7, glutamic acid;
8, glutamine; 9, glycine (+peptide); 10, histidine (+carosine);
11, leucine/isoleucine; 12, lysine; 13, methionine sulphoxide;
14, ornithine; 15, phenylalanine; 16, phosphoethanolamine
(+taurine); 17, proline ; 18, serine; 19, threonine; 20, tyrosine;
21, tyrosine-0-phosphate; 22, valine; 23-26, peptides; 27,
γ-aminobutyric acid; S, starting point. (From Chen et al(35).)

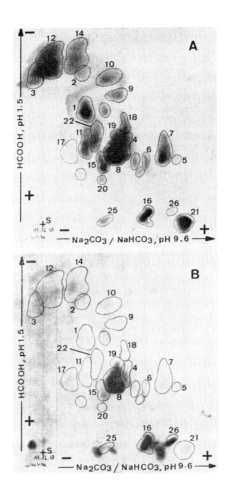

Fig. 9. Separation of amino acids, peptides and related compounds in the larval extract of the mutant l(3)tr of D.melanogaster by two-dimensional paper electrophoresis. Following an initial treatment with 0.5% ninhydrin (A), the electropherogram was sprayed with hypochlorite and the starch-iodide reagents (B). The numbering of the spots is the same as in Fig.8 except that no peptide is present in spots nos. 2, 9 and 10. (From Chen et al. (35).)

the peptide test revealed at least six spots which were either
invisible or colored very faintly by the ninhydrin reaction
(Fig 9 b). Furthermore, tyrosine phosphate, which is highly con-
centrated in the mature larvae of Drosophila, gave a heavy ninhydrin-positive spot, but failed to react with the hypochlorite-
starch-ioide reagent.

The method of Rydon and Smith (33) has several disadvantages:
First, the background is usualy somewhat blue. This problem can
be largely oververcome by repeatedly spraying the paper after
the hypochlorite step with mild reducing agents such as 96%
ethanol, 5% glucose or 1% formaldehyde. Second, the blue color is
not stable. But benzidine can be used instead of starch-iodide
(36), though the color developed is less intense. Third, the
reaction is not strictly specific. In addition to peptides, proteins, amides and many amino compounds give also positive reaction. Under the conditions used by us glutamine, methionine
sulphoxide and phosphoethanolamine were stained too (see Figs.
8 and 9). However, these will not be detected if their concentration in the smple is less than 0,025 - 0,05 M.

ION-EXCHANGE CHROMATOGRAPHY

The use of paper chromatography and paper electrophoresis for
amino acid analysis has the limitation that with samples which
do not overload the filter paper these methods allow the detection only of those components which occur at relatively high
concentrations. In this respect the technique of ionexchange
chromatography is superior. By a proper selection of the resin
and the eluting system an adequate separation can be achieved
even when a large quantity of the material in question is used.

1. General procedure

For work with amino acids, peptides and related compounds
sulfonated polystyrene resins in the Na^+- or H^+-form are most
commonly used (37). The resin is thoroughly washed and packed
in a column. Following equilibration with the initial buffer,
the sample to be fractionated is applied. The overall separation
depends on the particle size and the degree of crosslinkage

of the resin, the size of the column, the pH, ionic strength
and flowing rate of the eluting buffer, as well as the operation
temperature. With adequate control of these variables conditions
can be established which allow a complete separation of each
compound by the time it emerges from the column. The fractions
collected can then be analyzed by any one of the methods already
described.

One useful system based on the gradient elution procedure of
Thompson (38) has been successfully aplied to the study of amino
acids and derivatives in both Drosophila (24) and Culex (30).
As will be described below, the steps involved are more time-
consuming and less precise than those using the amino acid
analyzer. This system has, however, the advantage that the
fractions can be conveniently subjected to further analysis to
give additional information about the chemical composition of
the sample.

The procedures used by us are as follows: About 40 g of
Dowex 50-x4 (200/400) is shaken with 20 parts of 1 M NaOH and
allowed to stand for 4-6 h at room temperature. After removal
of the fine particles in the supernatant fraction, the resin is
washed on a sintered-glass funnel with distilled water until
neutral. Following this, the resin is suspended in 20 parts of
1 M HCl, allowed to settle for 4-6 h and again washed with
distilled water until HCl-free. The resin in the acidic from
is then slurried into a column of 1.0 x 40 cm in size and, after
settling, equilibrated with the starting buffer the pH of the
effluent is the same as that of the input solution.

A sample equivalent to about 2-3 g of wet weight is carefully applied to the column without disturbing the surface of
the resin, and collection of 5 ml-fractions started immediately.
Elution is effected with a gradient system using ammonium
formate and ammonium acetate in the buffer. At the beginning,
the mixing vessel is charged with 1 volume (53 ml) and the reservoir 2 volumes of 0.05 M ammonium formate. When all of the solution has run into the mixing vessel, the reservoir is charged
with 3 volumes of the next buffer in the sequence shown in Table 2.

In this way the pH of the eluting solvent increases from 2.5
to 8.0, and the concentration from 0.05 to 1.0 M (calculated
from ammonia). The acidic amino acids are eluted from the column first, while the basic ones emerge last. Under the conditions given here, aspartic acid appears usually in fractions
56-57, arginine in fractions 185-188, and the other amino acids
are between these.

2. Automatic recording system

In view of the laborious work involved in the analysis
of protein structures and the complex patterns of ninhydrin-
positive compounds in biological materials, the advantage of
automation of the whole operation is self-evident. In recent
years large progress has been made in this field, and various
types of amino acid analyzers are now commercially available.
The design of the apparatus is based on the method worked out
by Spackman, Stein and Moore (39). Instructions for the operation and maintenance of the equipment are usually given by the
manufacturer and thus require no further statement.

Table 2

Gradient buffer system for the fractionation of amino
acids and derivatives (From Chen (30).)

Buffer solution	pH	Molarity (NH_4^+)	Volume of buffer ml
1. Ammonium formate	2.50	0.05	106
2. Ammonium formate	2.90	0.10	159
3. Ammonium formate	3.30	0.15	159
4. Ammonium formate	3.65	0.20	159
5. Ammonium acetate	5.50	0.40	212
6. Ammonium acetate	6.80	0.60	159
7. Ammonium acetate	8.00	1.00	159

For our study of amino acids and peptides in insects we
used the Beckman/Spinco Amino Acid Analyzer 120 B and, more
recently, the Multichrom B. The latter model has several improvements: First, the improved quality of the resin allows the
use of a higher pressure to pump the buffer into the column.
Owing to the increase of the flow rate the time needed for a
complete analysis is much shortened. Second, in addition to
automation of both buffer and temperature changes, regeneration
and equilibration of the resin can also be programmed. Third,
by use of the automatic sample changer a maximum of 30 samples
can be analyzed in a single series. Fourth, by varying the absorption range the sensitivity of the colorimeter can be adjusted to the concentration of the components in the sample to
be analyzed. Acceptable accuracy can be obtained with a concentration as small as 10nM of each of the amino acids, though for
an optimal evaluation an average of about 50 nM per component
should be chosen.

Our previous conditions for operating the 120 B analyzer
are given in Chen and Hanimann (40). For use of the model Multichrom B the one-column   method , which is a recent development,
has been found to be most useful. This method allows the complete analysis of all ninhydrin-reacting components encountered
in the insect material by using only a single sample, and the
separation is accomplished within about 8 h. In our routine
analysis a sample equivalent to 0.05 g of larvae or flies is
injected into the sample changer. Depending on the final concentration of the extract the sample volume varies from 25 to 100
μl. Following setting of the fluid depth of the cuvette at 3 mm
and the absorption range of the colorimeter at 0-0.5 E, the
sample is run under the following conditions:

Column: 0.9 x 55 cm, resin M 72

Flow rate: buffer , 70 ml/h; ninhydrin, 35 ml/h

Eluting buffers:

     1. Sodium citrate, 0.2 N, pH 3.18, 140 min

     2. Sodium citrate, 0.2 N, pH 3.56, 135 min

     3. Sodium citrate, 0.2 N,/ NaCl, 0.6 N, pH 4.66, 190min

Temperature:

    1. 31°C, 123 min
    2. 55°C, 342 min

After each run the column is regenerated with 0.2 N NaOH for
10 min and equilibrated with the first buffer for 45 min.

As illustrated by the chromatogram in Fig. 10, all neutral
and basic amino acids are distinctly separated. But tyrosine
phosphate and glycerophosphoethanolamine emerge from the column
as a single peak, though they are at least partially separated
by use of the 120 B analyzer (Fig. 11). It should be possible
to improve the resolution of these acidic components by including an additional buffer of still lower pH in the eluting
system.

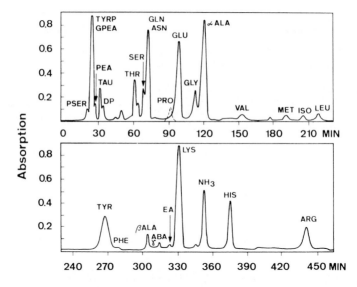

Fig. 10. Analysis of amino acids and derivatives in mature
larvae of D. melanogaster by automatic amino acid analyzer
(Multichrom B, Beckman). All abbreviations refer to the first
three letters of the corresponding amino acids with the following exceptions: PSER, phosphoserine; TYRP, tyrosine-0-phosphate;
GPEA, α-glycerophosphoethanolamine; PEA, phosphoethanolamine;
DP, dipeptide; GLN, glutamine; ASN, asparagine; ABA aminobutyric
acid; EA, ethanolamine.

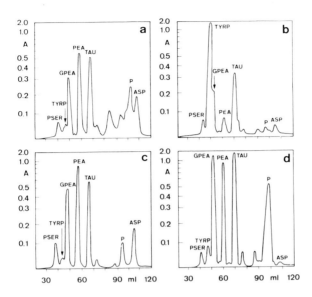

Fig. 11. Analysis of acidic components in methanolic extracts
from various stages of D. melanogaster by amino acid analyzer
(120 B, Beckman). a, 2-day-old larva; b, 4-day-old larva;
c, 2-day-old pupa; d, male fly aged 1 day after emergence. See
legend to Fig.10 for abbreviations. (From Chen and Hanimann
(40).)

SOME TECHNIQUES RELATED TO THE USE OF LABELLED AMINO ACIDS

As in other animals, for studies on the synthesis and turnover of proteins, peptides and related compounds in insects the
use of labelled amino acids is an indispensable tool. Various
methods have been devised to overcome the problems involved in
introducing the radioactive material into the egg, larva, pupa
or adult as well as in detecting its metabolic fate following
application. Some of these will be described in the following
sections. The general procedures for taking care of the radioactive amino acids and preparation of the sample for counting
can be found in special operation manuals devoted to isotope
techniques.

1. Isotope labelling of the developing egg

The most convenient way of labelling the insect egg is by
injecting the precursors into the adult female. This is an indirect method, as the injected material has to be transported
to the ovary through a rather complicated transport system. It
is known that yolk proteins in insects are synthesized in the
fat body, released to the hemolymph and then uptaken by the
oocytes. Thus the biochemical analysis of the developing egg
labelled in this way is a complex one.

More useful information can be obtained by direct application
of the radioactive material to the egg at the desired stage.
One handicap by using such a procedure is that the egg shell
allows the penetration of only non-polar molecules and compounds of low molecular weight (41). In terrestrial insects
air and water can move across the egg membranes. Thus, the egg
of Acheta domestica has been labelled by incubating it in an
atmosphere of $^{14}CO_2$ (42). The limit of this method is clear,
as it can not be used for such non-volatile precursors like
amino acids. An alternative procedure is injection, as has
been employed in studies of the coleopteran eggs (43, 44).
The difficulty here lies in the rupture of the egg membranes.
Since the insect egg usually has a high internal pressure and
its cytoplasm is fluid-like, loss of material at the site of
injection is inevitable, resulting in a disruption of the
spatial organization.

Sayles et al. (45) suggested the following procedure to
label Drosophila embryos: The chorion is first removed by a
brief treatment with 4.6% sodium hypochlorite. The dechorionated embryos are then suspended in Ringer's solution containing
0.5% Triton X-100 and the radioactive material. The adequacy
of this method is indicated by both high viability (over 90%)
and normal development of the labelled embryos. It is evident
that when following the proposed procedure the extent of dechorionation has to be examined carefully, and Ringer's solution
of sufficient ionic strength should be used to prevent diffusion of metabolites out of the embryo into the suspension
medium.

In their analysis of protein metabolism in the developing
locust egg topical application of labelled amino acids has been
used by McGregor and Loughton (46). Subsequent to dechorionation
with 1% hypochlorite, the egg was placed on a dried agar film
in a petri dish, and labelled leucine or arginine in 0.1 N HCl
or Ringer's solution was then applied to the egg surface. After
incubation at $37^{\circ}C$ for various length of time, the egg was subjected to analysis for radioactivity in both   the protein
fraction and the free amino acid pool. The results of these
authors showed that the uptake of the administered amino acids
through the dechorionated egg surface was adequate. One serious
problem here is desiccation. Thus, to ensure normal development
the humidity in the incubation dish has to be rigorously controlled.

2. Introduction of labelled amino acids into the larva, pupa
or adult

a. Feeding - In theory, labelled amino acids can be introduced into the larva or adult simply by adding these compounds
to the culturing medium. However, contamination of the diet
under usual laboratory conditions presents intractable problems.
More acceptable results can be obtained by raising the insects
aseptically on a defined medium containing the radioactive material. One inherent drawback of this method is the low
efficiency of incorporation. In our previous study on the
hemolymph proteins in Phormia, mature larvae were reared on a
synthetic diet consisting mainly of casein and ^{14}C-Chlorella
protein (47). Calculations from the average labelling and
number of larvae reared in each culture indicate that the incorporation amounts to no more than 10% of the total radioactivity added to the medium.

A more efficient method is to feed the larvae or flies on
an amino acid solution. Tyrosine-0-phosphate has been labelled
by keeping Drosophila larvae in a small volume of yeast suspension containing ^{14}C-tyrosine for about four hours (48). For
labelling adult flies of Drosophila we found that reproducible
results can be obtained by the following procedure: About 25

flies are placed in a 25 ml-Erlenmeyer flask and kept without
food for 1 h. Following this, a drop (10 µl) of 25% sucrose
solution containing the labelled amino acids is pipetted into
the flask. A separate drop of water is also pipetted into the
flask to reduce evaporation of the sugar solution. Usually the
flies will finish drinking the offered solution within 1-2 h.
Control experiments showed that less than 1% of the total
radioactivity was left in the flask. In the procedure used by
Clarke and Maynard-Smith (49) flies were allowed to drink a
10% sucrose solution containing tritiated leucine from a fine
glass pipette. The amount of solution consumed was estimated
by weighing the flies before and after drinking.

   b. Injection - All methods of administering the labelled
amino acids by feeding suffer from the disadvantage that the
compounds have to pass through the digestive tract. An alternative method is to inject the precursors directly into the
hemocoel. The injection is generally performed by employing
a calibrated glass needle attached to a metal holder, which in
turn is connected to a microinjection apparatus with a piece
of capillary Teflon tubing. By use of an Agla micrometer syringe (Wellcome Research Laboratory,Beckenham, Fngland) we
found that about 1-2 µl solution can be injected into a larva
or fly of the size of Phormia with reasonable accuracy (47).

   When a large number of animals have to be injected, the use
of the type of equipment mentioned above becomes rather tedious
because of the need for continuous reading of the micrometer
and refilling of the glass needle. For serial injections of
Drosophila larvae Mitchell (50) developed an apparatus with an
injection volume of as little as 0.05 µl. A similar type of
injection apparatus has been constructed by us (51, 52). The
diagram in Fig. 12, which illustrates the design of the apparatus, is self-explanatory. Its operation is very simple. Before injection the glass tube (i) and the attached Teflon
capillary tubing (k) are filled with water. A fine glass needle
(1) is then inserted into the other end of the Teflon tubing
which carries a holder with a holding clamp (not shown in the
diagram) to ensure tight connection. Following this, the

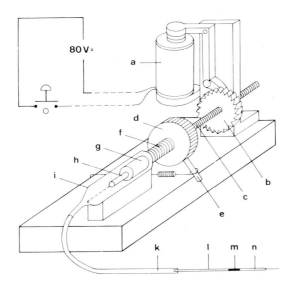

Fig. 12. Diagram of apparatus for serial injections. See text
for operation of the apparatus . (From P. Baumann (52).)

rotating wheel (d) is turned backwards to fill the needle first
with a tiny drop of paraffin oil (m) and then the isotope solution (n) to be injected. By pushing the pedal switch (o) the
steel axis (h) in the glass tube moves forwards to deliver a
volume of 0.037 µl. The accuracy of the injected volumes is
approximately ± 20%.

In practice, groups of about 20 larvae or flies are etheri-
zed and stuck parallel to each other onto a piece of adhesive
tape mounted on a slide. These are then injected under a
binocular dissecting microscope with convenient magnification.
During injection the animal is gently held down with a dissecting needle in the left hand, and the glass needle inserted
into the body cavity by keeping the holder in the right hand.
The desired volume of solution is then injected by pressing
briefly the switch with the foot. With some experience a large
number of injections can be performed within a short time, and
the mortality of the injected animals is virtually negligible.

Injection of the pupae is handicapped by the presence of a hard pupal case. In order to overcome this difficulty, one can pierce the pupal case with a fine tungsten needle and then insert the glass needle into the aperture for injection. If necessary, the wound is sealed with a drop of paraffin wax.

3. Recording of $^{14}CO_2$ production

The measurement of $^{14}CO_2$ production provides a useful criterion for judging the extent of oxidation of the injected amino acid. Various methods have been devised for a continuous recording of the expired $^{14}CO_2$ following injection. In the procedure used by Robinson and Chefurka (53) $^{14}CO_2$ in the metabolism cage produced by insects injected with labelled glucose was pumped to circulate through an ionization chamber and the radioactivity recorded by a vibrating-reed electrometer. Levenbook and Dinamarca (54) simplified the above procedure in the following way: An adult blowfly injected with either ^{14}C-alanine or ^{14}C-lysine is placed in a cylindrical wire-gauze cage attached to the end of a threaded brass plug (Fig.13). The latter is then screwed tightly into a Cary-Tolbert flow-through ionization chamber for recording with a Cary vibrating-reed electrometer. Since the cage carrying the labelled fly is directly placed in the chamber, the sensivity of the simplified procedure (1 mV = 2.7 x 10^{-4} C) is about three times higher than that reported previously by Robinson and Chefurka (53).

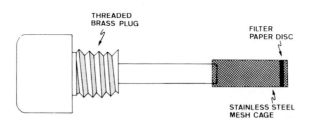

Fig. 13. Diagram of brass plug and insect cage for recording respiratory $^{14}CO_2$ in ionization chamber. (From Levenbook and Dinamarca (54).)

Another method has been developed in our laboratory for recording $^{14}CO_2$ formation in adult Drosophila. An aluminum chamber with a volume fo about 5 ml is mounted tightly in front of the window of a Nuclear Chicago D34 Geiger-Muller tube (Fig. 14). A Wire-gauze cage containing the fly is provided with an aluminum shield on its side towards the window of the tube and placed in the chamber. The radioactivity of the $^{14}CO_2$ produced can then be measured by an appropriate counting system. The sensitivity has been estimated to be 6800 cpm/ C.

For accurate measurements the apparatus has to be calibrated by comparing the values obtained with those calculated from parallel samples by assuming that loss of radioactivity, exepting that by excretion, is all due to oxidation. Injection of ^{14}C-alanine into adult males of Drosophila aged 3-50 days showed that the radioactivity of $^{14}CO_2$ accumulated at a 2 h-period amounted to 50-65% of the injected dose, which is within the range previously reported by Dinamarca and Levenbook (55).

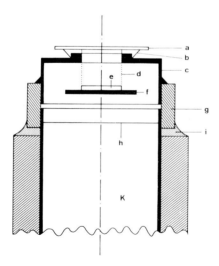

Fig. 14. Diagram of apparatus for recording respiratory $^{14}CO_2$ with a Geiger-Muller tube. a, cover glass; b, vaseline; c, aluminum chamber; d, wire-gauze cage; e, moist filter paper; f aluminum schield; g, plastic ring; h, window of G-M tube; i, paraffin wax; k, Nuclear Chicago D34 G-M tube. (From P. Baumann (52).)

ISOLATION AND PURIFICATION OF PEPTIDES AND RELATED COMPOUNDS

For the isolation of peptides and similar compounds from
the large pool of ninhydrin-positive components in insects
ion-exchange chromatography and gel filtration have been shown
to be most useful. The general procedure of ion-exchange chromatography has been described in a previous section. For preparative purpose the same operation conditions can be followed.
But care has to be taken so that there is no overloading of
the column by using a large sample.

For gel filtration we found that Sephadex (Pharmacia, Upp-
sala, Sweden) is most suitable for work with insect material.
It is a cross-linked dextran gel which separates the components
on the basis of a molecular sieve mechanism. Two properties of
the Sephadex gel are of crucial importance for its performance:
the particle size and the swelling ability. Smaller particles
have a more efficient separation and thus a higher resolving
power. For preparative work gels of medium or coarse particle
size are more convenient, as these allow the use of low operating pressure to maintain an efficient flow rate. The swelling
property of the gel sets a limit to the molecular weight range
of the substances to be separated. Sephadex preparations of
different particle sizes and degrees of swelling are commercially available.

A gel column is prepared by suspending the dry product in
an excess of dilute effluent buffer. We used to leave the suspension to stand overnight at room temperature in order to
allow the gel to swell sufficiently. Following deaeration, the
gel slurry is poured into a chromatographic tube. Before use
the column is washed with the solvent and the flow rate adjusted to give proper separation.

In connection with our studies on the paragonial secretions
in male adults of Drosophila, various ninhydin-positive compounds have been isolated in the author's laboratory. A polypeptide, PS-1, and a glycine derivative, PS-2, were isolated
from D. funebris (56). In a typical experiment the isolation
procedures were carried out as follows: The methanolic extract

from 30 g of male flies was applied to a Sephadex G-50 column
(2 x 64 cm) and eluted with 0.5 M acetic acid. A potion (0.3%)
of each of the fractions collected was subjected to analysis
for ninhydrin-positive components, and those containing the
polypeptide or glycine derivative were pooled as indicated in
Fig. 15.

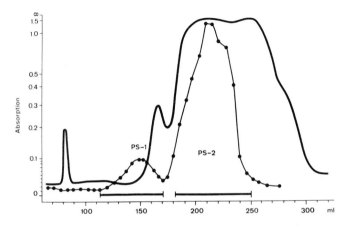

Fig. 15. Separation on Sephadex G-50 of a polypeptide (PS-1)
and a glycine derivative (PS-2) in methanolic extract from
30 g male flies of D. funebris. (From H. Baumann (56).)

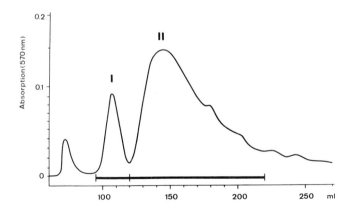

Fig. 16. Purification of polypeptide PS-1 from D. funebris on
Dowex 50W-x2. (From H. Baumann (56).)

The pooled polypeptide sample was evaporated in vacuo, dissolved in 10% acetic acid and again fractionated on a Dowex
50W-x2 column (1.6 x 66 cm) with 0.3 M pyridine acetate (pH
4.3). During fractionation 10% of the effluent was analyzed on
an amino acid analyzer by making use of the stream-dividing
system. The elution profile is shown in Fig. 16. Fingerprint
analysis revealed that both peaks I and II contained the polypeptide, but peak II was contaminated with phosphorylated amino
acids. The contaminations and the pyridine buffer in the pooled
fractions were removed on Sephadex G-25 in 0.1 M acetic acid.
This procedure yielded 24mg of peptide with a total recovery
of 80-90%.

Analysis of the amino acid sequence demonstrated that the
isolated product contains two peptide chains each with 27 amino
acid residues (57).The two chains differ only in valine and
leucine in the second position, and the ratio of the valine to
the leucine chain is 7:3. Injection of the purified peptide
into virgin females resulted in reduction of their receptivity
to copulation.

The pooled glycine derivative from Sephadex G-50 chromatography was also concentrated and chromatographed again on
Dowex 50W-x8 equilibrated with 0.3 M pyridine acetate. It
emerged from the column between γ-aminobutyric acid and ethanol-

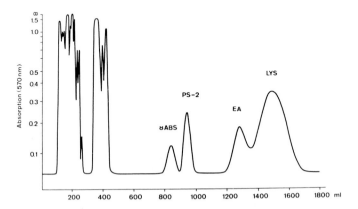

Fig. 17. Isolation of the glycine derivative PS-2 from
D. funebris on Dowex 50W-x8. (From H. Baumann (56).)

amine (Fig. 17). Following partial removal of the pyridine
buffer on Dowex 2-x8 a final purification step was performed
on a column of Sephadex G-10 fine in 0.1 M acetic acid. The
total yield was 9 mg, and the overall recovery about 50%.

Low-molecular-weight peptides have been isolated from early
3rd instar larvae of the blowfly Phormia regina by Levenbook
(58, 59). The larval extract, deproteinized with 3% sulpho-
salicylic acid, was desalted on Dowex 50 with 4 M NH_3. After
evaporation of the ammonia, the residue was dissolved in 0.01 M
collidine acetate buffer, pH 8, in an excess of cupric carbonate. The copper complexes of amino acids and peptides were
separated into five fractions on a DEAE-cellulose column.
Subsequent to removal of Cu^{2+}, further separation of each
fraction was effected by preparative electrophoresis. Among a
total of 9 peptides isolated in this way 8 were found to be
dipeptides and one was a tripeptide. All of them have either
histidine or lysine as the N-terminal amino acid.

A dipeptide, β-alanyl-L-tyrosine, has been isolated form
the larvae of the fleshfly Sarcophaga bullata by chromatography
on Dowex 50-x12 in 0.51 M sodium citrate-HCl buffer, pH 4.25
(60), and another dipeptide, γ-L-glutamyl-phenylalanine, from
the larvae of the housefly Musca domestica by chromatography on
Dowex 2 with water and 0.2-1.0 M acetic acid as solvents(61).
Both compounds serve as tyrosine reservoir for the scleroti-
zation of the cuticular proteins during puparium formation.

In Drosophila melanogaster tyrosine required in the tanning
reaction is stored in the form of a phenol phosphate ester,
tyrosine-0-phosphate (48,62). The crude extract from mature
larvae was applied on a column of Dowex 1-x8 in the acetate
form. Elution was first performed with water to remove most
basic and neutral amino acids and followed by 0.2 to o.5 M acetic
acid. Those fractions which contained the tyrosine phosphate
were combined, and concentrated, and the isolated compounds was
repeatedly crystallized.

In general, for the isolation of large peptides it is
advantageous to do a partial purification on Sephadex first,

as this will increase the ease of subsequent separation by ion-
exchange chromatography. For work with compounds of low molecular weight this may not be necessary. When the isolated
product has to be injected into larvae or flies to test its
biological activity, a final purification step by gel filtration
is in any case recommended. In our experience traces of ammonia
or other contaminating materials in the solution are usually
very toxic.

CONCLUDING REMARKS

   Rapid progress has been made in the study of amino acids,
peptides and related compounds in insects by taking advantage
of the recent development of adequate methods. There are,
however, several inherent difficulties which have to be
considered seriously in the design of experiments. First, the
patterns of amino acids and their derivatives are highly
species-specific. It is not possible to extrapolate the results
obtained from one insect to even closely related species.
Second, the occurrence and concentration of these components
vary considerably in the course of development and aging.
Thus, the morphogenetic state and age of the animals used
should be rigorously controlled. Third, the nutritional
condition has a large influence on the biochemical make-up
of the insects.A meaningful interpretation of the results is
possible only when the animals are raised under carefully
controlled conditions.

   In the present chapter only procedures of some selected
chromatographic and electrophoretic methods have been described.
The choice is to a large extent determined by the author's own
experience. Needless to say, many other techniques such as
thin layer chromatography, immunoelectrophoresis and micro-
biological assay, are equally useful. No attempt has been
made to list all references in which the methods described in
the foregoing sections have been used. A detailed description
of the biochemical properties of amino acids has also been
omitted, as this would be beyond the scope of this article. In
spite of these limitations, it is hoped that this chapter will

help those who are interested in insect amino acids to choose
the most promising methods to solve their particular problems.

ACKNOWLEDGEMENTS

The original work reported in this article was supported
by the Swiss National Science Foundation, the Georges and
Antoine Claraz-Schenkung and the Karl Hescheler-Stiftung.

REFERENCES

1   J.L. Auclair, Can. Ent., 85(1953)63
2   M. Florkin, in L. Levenbook (Editor), Proc. 4th. int.
    Congr. Biochem., Pergamon Press, London, 1959, p.63.
3   J.B. Buck, in K.D. Roeder (Editor), Insect Physiology,
    Wiley, New York, 1953, Ch.7, p.191.
4   P.S. Chen, Adv. Insect Physiol., 3(1966)53.
5   P.S. Chen, Biochemical Aspects of Insect Development,
    Karger, Basel, 1971.
6   P.S. Chen, in J.T. Holden (Editor), Amino Acid Pools,
    Elsevier, Amsterdam, 1962, Ch.4, p.115.
7   J.L. Auclair and R.L. Patton, Rev. can. Biol., 9(1950)3.
8   J.J. Corrigan and N.G. Srinivasan, Biochemistry 5 (1966)1185.
9   N.G. Srinivasan, J.J. Corrigan, and A. Meister, J. biol.
    Chem., 240(1965)796.
10 D.R. Rajagopal, A.H. Ennor and B. Thorpe, Comp. Biochem.
    Physiol., 21(1967)709.
11 F. Irreverre and L. Levenbook, Biochim. Biophys. Acta,
    38(1960)358.
12 R. Pant and H.C. Agrawal, J. Insect Physiol., 10(1964)443.
13 J.L. Villeneuve, J. Insect Physiol., 8(1962)585.
14 D.I. Wang and F.E. Moeller, J. Invert. Pathol., 15(1970)202.
15 E. Lederer and M. Lederer, Chromatography, Elsevier,
    Amsterdam, 1953.

16   F. Cramer, Papierchromatographie, Verlag Chemie, Weinheim, 1962.

17   R. Clotten and A. Clotten, Hochspannungselektrophorese, Thieme, Stuttgart, 1962.

18   H. Determann, Gelchromatographie, Springer, Berlin, 1967.

19   J.L. Bailey, Techniques in Protein Chemistry, Elsevier, Amsterdam, 1967.

20   i. Smith, Chromatographic and Electrophoretic Techniques, Heinemann, London, 1968.

21   G. Zweig and J.R. Whitaker, Paper Chromatograhy and Electrophoresis, Academic Press, New York, 1967.

22   S. Blackburn, Amino Acid Determination, Dekker, New York, 1968

23   K. Dorfner, Ionenaustauscher, de Gruyter, Berlin, 1970.

24   H.K. Mitchell and J.R. Simmons, in J.T. Holden (Editor), Amino Acid Pools, Elsevier, Amsterdam, 1962, Ch.4, p.136.

25   J.L. Auclair and R.Dubreuil, Can. J. Zool., 30(1952)109.

26   E. Hadorn, Developmental Genetics and Lethal Factors, Methuen, London, 1961, p.260.

27   F.G. Fischer and H. Dörfel, Biochem. Z., 324(1953)544.

28   P.S. Chen and C. Diem, J. Insect Physiol., 7(1961)289.

29   P.S. Chen, F. Hanimann and C. Roeder, Rev. suisse Zool., 73(1966)219.

30   P.S. Chen, J. Insect Physiol., 9(1963)453.

31   L. Levenbook, Comp. Biochem. Physiol., 18(1966) 341.

32   D. Gross, Nature, 184(1959)1298.

33   H.N. Rydon and P.W.G. Smith, Nature, 169(1952)922.

34   S.C. Pan and J.D. Dutcher, Anal. Chem., 28(1956)836.

35   P.S. Chen, E. Kubli and F. Hanimann, Rev. suisse Zool., 75(1968)509.

36   F. Reindel and W. Hoppe, Naturwiss., 40(1953)221.

37   S. Moore, D.H. Spackman and W.H. Stein, Anal. Chem., 30(1958) 1185.

38   A.R. Thompson, Biochem. J., 61(1955)253.

39   D.H. Spackman, W.H. Stein and S. Moore, Anal. Chem., 30(1958)1190.

40 P.S. Chen and F. Hanimann, Z. Naturforsch., 20b(1965)307.

41 T.O. Browning, J. exp. Biol., 56(1972) 769.

42 E. Hansen-Delkeskamp, H.W Sauer and F. Duspiva, Z. Naturforsch., 22b(1967)540.

43 R.A. Lockshin, Science, 154(1966)775.

44 H.W. Küthe, Wilhelm Roux'Arch., 172(1973)58.

45 C.D. Sayles, J.D. Procunier and L.W. Browder, Nature-New Biol., 241(1973)215.

46 D.A. McGregor and B.G. Loughton, Wilhelm Roux' Arch., 179(1976)

47 P.S. Chen and L. Levenbook, J. Insect Physiol. 12(1966)1611.

48 K.D. Lunan and H.K. Mitchell, Arch. Biochem. Biophys., 132(1969)450.

49 J.M. Clarke and J. Maynard-Smith, Nature, 209(1966)627.

50 J.R. Simmons and H.K. Mitchell, in J. Holden (Editor), Amino Acid Pools, Elsevier, Amsterdam, 1962,Ch.4, p.147.

51 P.A. Baumann and P.S. Chen, Rev. suisse Zool., 75(1968)1051.

52 P.A. Baumann, Z. vergl. Physiol., 64(1969)212.

53 J.R. Robinson and W. Chefurka, Anal. Biochem., 9(1964)197.

54 L.Levenbook and M.L. Dinamarca , Anal. Biochem., 11(1965)391.

55 M.L. Dinamarca and L. Levenbook, Arch. Biochem. Biophys., 117(1966)110.

56 H. Baumann, J. Insect Physiol., 20 (1974) 2181.

57 H. Baumann, K.J. Wilson, P.S. Chen and R.E. Humbel, Eur. J. Biochem., 52(1975)521.

58 L. Levenbook, Comp. Biochem. Physiol., 18(1966)341.

59 R.F. Bodnaryk and L. Levenbook, Biochem. J. 1lo(1968)771.

60 L. Levenbook, R.P. Bodnaryk and T.F. Spande, J. Biochem., 113(1969)837.

61 R.F. Bodnaryk, J. Insect Physiol., 16(1970)919.

62 H.K. Mitchell and K.D. Lunan, Arch. Biochem. Biophys., 106(1964)219.

63 H. Baumann and P.S. Chen, Rev. suisse Zool., 80(1973)685.

64 P.S. Chen and A. Oechslin, J. Insect Physiol. (in press).

65 P.S. Chen, Rev. suisse Zool., 82(1975)673.

CHAPTER 5

INSECT LIPID ANALYSIS

Larry L. Jackson and Melvin T. Armold

Department of Chemistry,Montana State University,Bozeman, Montana 59715, U.S.A.

CONTENTS

I. INTRODUCTION

Insect lipids function as essential and integral components of cell membranes and as important sources of metabolic energy for cell mainten- ance, flight, reproduction, embryogenesis, and metamorphosis. They also serve communicative roles as pheromones and kairomones, regulative roles as hormones, and protective roles as cuticular lipids, beeswax and scale secretions. The lipids of only a few of the many insect species have been examined to date. While generalizations are certainly possible, it is ob- vious that insects have considerable variety in lipid composition and spec- ific function. Furthermore the investigators of insect lipids have used a wide variety of techniques which complicate direct comparison of results

from their studies. This chapter is an attempt to compile techniques used
in insect lipid analysis and to suggest new lipid methodologies which have
not yet been applied to insect lipid analysis.

We have structured this chapter to provide in the first three sections
information of general interest to the lipid researcher. The second three
sections cover the most extensively studied areas of insect lipid research,
fatty acids, sterols and polar lipid analyses. In the remaining sections,
more specialized studies of lipids from specific tissues and groups of insects are presented. Hydrocarbon analysis is included under cuticular lipids and beeswax, wax ester analysis under beeswax, scale insects and
cuticular lipids, and fatty acid methyl ester analysis under insect eggs.

II. PREPARATIONS AND PRECAUTIONS

The first procedure used in the analysis of lipids is a precautionary
one. All glassware and other equipment which will contact the lipid samples must be scrupulously clean. Clean glassware will shed water equally
well over its entire surface. Oil or grease spots on the glassware are seen
as regions where the water doesn't flow as a sheet or as regions where
droplets of water form. Previously used and uncleaned glassware and fittings
involved in rotary evaporators, and glass tubing used to evaporate under
nitrogen are potential sources of contamination. The use of rubber or
plastic tubing must be avoided at all times since plasticisers and other
materials can be extracted by organic solvents. The lipid researcher must
be aware of potential sources of contamination.

The solvents used during the analysis should be examined for purity.
This can be accomplished by evaporating a quantity larger than the amount
which will normally be used in the analysis. The absence of any visible
residue indicates purity; however a small quantity of fresh solvent should
be added to the residue or suspected residue and analyzed by thin layer
chromatography (TLC) and gas liquid chromatography (GLC). The total absence of spots on the TLC plate or peaks other than solvent in the gas chromatogram along with no visible residue argues that this lot of this particular solvent is pure. Any solvents which are not pure must be redistilled before use. It is often advantageous to simply redistill all solvents used
routinely since this eliminates the necessity of checking each new lot used.

These precautionary measures may seem overstated to some readers,
but contamination of the lipid samples during analysis is a potentially
serious problem. There are reports of substances subsequently found to be

artifacts (1). The proper precautions go a long way towards eliminating problems with artifacts.

III. EXTRACTION TECHNIQUES

Homogenization of whole insects or insect tissue has been carried out in a number of ways generally dependent upon the amount and hardness of the material to be homogenized. For softer tissues the Potter–Evehjem or some form of all-glass homogenizers (Ten Broeck) are frequently used; whereas for harder materials and larger tissue volumes, homogenizers with knives or other sharp cutting edges are used. In many techniques the extracting solvent is the solvent used for homogenization, but other techniques use a homogenization medium which is then extracted with solvent.

Although many solvent systems have been used for insect lipid extraction, by far the most popular protocol is that of Folch, Lees and Sloan-Stanley (2) which was used by Vroman, et al. (3), Fast (4), Kinsella and Smyth (5), Wimer and Lumb (6), Beenakers and Gilbert (7), Cmelik (8), Harlow, et al. (9), Kinsella (10), Bridges and Price (11), Beenakkersand Gilbert (12), Cenedella (13), Chippendale (14), Cohen (15), Dutkowski and Ziajka (16), Lambremont (17), Beenakkers (18), Brown and Chippendale (19) Tan (20) and many others. The Folch et al. (2) technique has been modified by many researchers and utilized for insect analysis. Bieber et al. (21) made modifications that were used by Sun and Brookes (22). D'Coste and Birt (23) made modifications that were used by Walker et al. (24). In the same year, Lambremont et al. (25) made modifications adopted by Mauldin et al. (26). Schaefer (27) used 3 volumes of chloroform to 2 volumes of methanol. Overturf and Dryer (28) used 0.29 percent NaCl as the wash (29). Stevenson (30) used a water:methanol (1:1 v/v) wash of the filtered extract to remove non-lipid material as suggested by Radin (31). Thomas (32) re-extracted the residue with chloroform:methanol saturated with ammonia (7:1 v/v) as suggested by Rouser et al. (33).

Another frequently used extraction technique is that of Bligh and Dyer (34) which also utilizes chloroform and methanol but uses less solvent and a slightly different approach than the Folch et al. (2) technique. The Bligh and Dyer method has been recently used by Barlow (35), Castillon et al. (36), Thompson and Bennett (37), Thompson and Barlow (38) and Dutkowski et al. (39).

An ether:methanol (3:1 v/v) mixture was used by Hutchins and Martin (40). Dutkowski and Ziajka (16) used ethanol:petroleum ether (3:1 v/v) as the extracting solvent. The so called Dole's solvent (isopropanol:heptane: $1\underline{N}\,H_2SO_4$ 40:10:1 v/v/v) (41) was used by Chang and Freidman (42). Madariaga et al. (43) used the rather interesting combination of tissue homogenization in 8.5 percent sucrose followed by filtration, then addition of chloroform and methanol to the filtrate to form a single phase. After stirring under nitrogen, chloroform:methanol (1:1 v/v) was added to form two phases which were then centrifuged and separated. Chloroform was used to re-extract and 0.75 percent NaCl was used to wash the organic layer. An alternative to washing any of the above lipid extracts to remove non-lipid contaminants is to use Sephadex LH-20 (44). This technique is particularly suggested when "interface fluff" is a serious problem.

The soxhlet apparatus utilizing light petroleum or methylene chloride (45,46) and chloroform (47) has been used recently for insect lipid extraction but is not as popular as it is in other areas of lipid research. Micromethods for extracting small quantities of tissue such as that of Kaschnitz et al. (48) may find considerable use in insect lipid analysis in the future.

The quantity of extracted lipid can be expressed in any of several conventions. A convention commonly used in the older literature is the expression of the mass of the extracted lipid per mass of tissue or organisms used. Although frequently used this procedure is often criticized as unsatisfactory because the mass of the tissue or organism will vary with the degree of hydration of the organism. A second convention, the percentage lipid of the lean dry weight, is now frequently used (49). For this convention the lipid is extracted, the solvent is removed, and the extract is weighed. The extracted tissue is then dried and weighed. The ratio of extracted lipid to lean dry weight is then simply expressed as a percentage.

Presently some workers present their data as both percentages, the % wet or fresh weight to facilitate comparison with the older literature and the % lean dry weight to provide presumably more reliable extraction data. A third convention which may be encountered is the expression of the mass of extracted lipid per organism. This helps visualize the relationship of lipid to the individual insect, but again is presumably less accurate since the mass of the individuals will vary, and the degree of hydration may also vary. The percent lean dry weight is the most common convention used presently since it yields the least variable results of the methods commonly used (49).

It is evident that a multitude of extraction techniques exist and space limitations prevent a comprehensive treatment. Radin (31) has compiled a number of lipid analysis techniques.

IV. CHROMATOGRAPHIC SEPARATIONS

After homogenization, extraction and removal of most or all non-lipid contaminants, the separation of the extracted lipid can begin. Adsorption chromatography using silicic acid (silica gel) or Florisil is commonly used since relatively large quantities of lipid can be separated simply by increasing the amount of adsorbent used. Many researchers prefer to make a separation of neutral lipids and polar lipids (mostly phospholipids) prior to further separations. Many of these procedures are based on the technique of Borgstrom (50). Marks et al. (51) mixed the lipid extract in anhydrous ether with silicic acid, then centrifuged and washed three times with ether to obtain the neutral lipids. The phospholipids were removed by washing the silicic acid with methanol. This technique has been used by Cenedella (13), and Brown and Chippendale (19, 52). Sun and Brookes (22) used the method of McCarthy and Duthrie (53) which involves adsorbing the lipid in ether onto a column of silicic acid and then eluting neutral lipids with ether, eluting free fatty acids with 2 percent formic acid in ether (v/v) and eluting phospholipids with methanol. Chang and Sweeley (54), Wood and Snyder (55) and Walker et al. (24) developed systems for separating neutral lipids and phospholipids which involved eluting the lipid laden silicic acid with chloroform to obtain the neutral lipids and eluting with methanol to obtain phospholipids. These techniques were adopted for insect lipid analysis by Wimer and Lumb (6), Harlow et al. (9), Mauldin et al. (26) and Lambremont (17).

The neutral lipids obtained from either the whole extract or the above preliminary separation are often chromatographed on silicic acid by the method of Hirsch and Ahrens (56) which uses increasing amounts of diethyl ether in petroleum ether or hexane. Kinsella and Smyth (5) have utilized this technique on insect lipids. The use of Florisil as recommended by Carroll (57,58) has been quite popular with insect researchers (7,12,22,30, 32,40). Alumina has not been used extensively but Cavill et al. (46) used it to separate hydrocarbons and glycerides, and Durocher and Leboeuf used it to separate long chain fatty acids from neutral lipids (59).

A few precautionary comments are in order concerning the technique of column chromatography. Since the polarity of the solvents determines the elution rate of the lipids from the column, it is important that the lipids be applied to the column as a solution (or occasionally, if necessary, as a partial suspension) in a minimum volume of the least polar solvent of the elution series. If too much solvent or the improper solvent is used, the bands will be too broad and resolution will be lost. It is also important to note that although reproducible results are obtainable, one must never assume that the eluted fractions from one column run will be identical to those obtained from another run. The fractions should always be analyzed by TLC for purity and for comparison to standards. The use of continuous monitoring of the column eluate by flame ionization detectors (60) and recording diffractometers (61) has been limited by difficulties with detectors. The development of new and better detectors for high pressure liquid chromatography may provide a new generation of detectors applicable to continuous monitoring.

If column chromatography yields fractions which contain mixtures of lipid classes, repeated column chromatography or the use of preparative TLC is recommended. Before preparative TLC, the mixture to be separated should be analyzed by analytical TLC using one or more solvent systems until complete separation of the lipid classes is obtained. Then a quantity of lipid (usually 25 mg for 0.25 mm thick plates and 25-100 mg for 0.50 mm thick plates) is applied as a narrow band to the plate. If the plate has not been overloaded, the separation will resemble or duplicate the separation obtained by analytical TLC. Beenakkers and Gilbert (7) and Thomas (32) used silica gel coated plates developed in hexane:ether: acetic acid(80:20: 2 v/v/v) or benzene:ethyl acetate (2:1 v/v) and visualized the bands with iodine vapor, or Rhodamine 6-G under U.V. light. Following separation of the bands, the lipid is extracted from the silica gel with chloroform:methanol:water (65:40:10 v/v/v) according to Privett et al.(62). Wimer and Lumb (6) used a hexane:ether:acetic acid (90:10:1 v/v/v) solvent system to develop the plates and extracted the silica gel with chloroform:methanol (2:1 v/v) and methanol. Cohen (15) developed the plates three times in petroleum ether:ether:acetic acid (85:15:1 v/v/v), visualized with iodine vapor and extracted with chloroform by centrifugation and decanting.

Much of the analytical TLC of insect neutral lipids is based on the early paper by Mangold (63). Insect researchers (5,14,64) have used silica gel plates developed in n-hexane:ether:acetic acid (90:10:1 v/v/v) to separate neutral lipids into classes. Peled and Tietz (65) prefer the system of Brown and Johnston (66) who used n-hexane:ether:acetic acid:methanol (90:20:2:3 v/v). Cmelik (8) found that petroleum ether:ether:acetic acid (85:15:2 v/v/v) provided separation of insect lipids. In general the use of varying ratios of hexane or petroleum ether to ether will provide good separations for non-polar insect lipids. Acidification of the solvent with a small quantity of acetic acid will mobilize the free fatty acids. A solvent system of ether:2,2,4-trimethylpentane (80:20 v/v) was used by Chippendale (64) to separate monoglycerides, diglycerides and sterols. The Freeman and West (67) system of double-development of the plate first in ether benzene:ethanol:acetic acid (40:50:2:0.2 v/v) followed by hexane:ether (94:6 v/v) was used on insect lipids by Tan (20) and Chang (29). The double-development technique of Hojnacki and Smith (68) also provides separation of neutral lipids on one chromatoplate.

Nondestructive visualization of the spots is usually done by exposure to iodine vapors or by spraying the plate with Rhodamine 6-G in ethanol or 2',7'-dichlorofluorescein in ethanol followed by viewing under ultraviolet light. A more permanent record can be made by spraying the plate with a saturated solution of potassium dichromate in 80% sulfuric acid by weight (67,69) followed in each case by heating at 120-150 C for 15 min to 25 min or more. For quantitation by densitometry Kinsella (10) and Brown and Chippendale (19) used the method of Privett et al. (62). The method of Fewster et al. (70) which involves spraying with 3% cupric acetate in 8% aqueous phosphoric acid and then heating at 180 C for 25 min was used by Thomas (32).

V. FATTY ACID ANALYSIS

Some insect lipid researchers have made estimates of total lipids, glycerides and free fatty acids based on various micro determinations. Stevenson (30) used the Van Handel and Zilbersmit (71) method for glyceride determination. The reduction of acidified potassium dichromate (72) was used by Walker et al. (24) to estimate lipids and total fatty acids. The method of Dole and Meinertz (41) for fatty acid determination is in common use among insect lipid researchers (16,30,39). Peled and Tietz (65) preferred the colorimetric technique of Duncombe (73).

Many investigations of insect lipids are primarily concerned with analysis of total fatty acids, glyceride fatty acids, phospholipid fatty acids and free fatty acids. In general these analyses involve 4 steps: 1) Saponification followed by methylation, or transesterification, 2) GLC of the fatty acid methyl esters, 3) Identification of the fatty acid methyl ester peaks, and 4) Quantitation of each fatty acid. The next few paragraphs review the various techniques used in fatty acid analysis. The review is not exhaustive but a number of papers were selected from techniques used around the world.

1) Saponification-methylation, and transesterification.

Saponification of insect lipids has been accomplished in a variety of solutions. Lindsay and Barlow (74) used 0.5N KOH in 95% ethanol at 90C for 1 hour, and Schaefer (24) used 5% ethanolic KOH under reflux for 3 hours with a nitrogen atmosphere. After the saponification reaction the mixture must be acidified, usually with acetic acid or dilute mineral acids, and extracted with hexane, heptane, petroleum ether or diethyl ether.

Methylation using BF_3-methanol according to the procedure of Metcalfe and Schmitz (75) or the modified procedure employing saponification and methylation (76) are popular methylation procedures for insect lipids (27,32, 40,77). A number of modifications have been made in the Metcalfe and Schmitz procedure, for example Hyun et al. (78) did transesterification in 4 parts of BF_3-methanol and 1 part of benzene, followed by extraction with petroleum ether. This procedure was used on insects by Cenedella (13). Morrison and Smith (79) modified the use of BF_3- methanol to include an extraction of the methyl esters with pentane, and found BF_3 was better than BCl_3. Martin (80) used the Morrison and Smith procedure on insects. The use of heptane for extraction of the reaction mixture of Metcalfe and Sch- mitz to improve quantitation was suggested by Van Wijunaarden (81). Bee- nakkersand Gilbert (7) extracted with hexane. Peterson et al. (82) used BCl_3-methanol and found it more active in transesterification, providing faster rates. Beenakkers and Gilbert (7) modified the Morrison and Smith (79) procedure by using a mixture of benzene:BF_3 in methanol (14 percent w/v):methanol (1:1:1) and used the procedure on a number of insects (12, 18).

Methylation with BF_3-methanol has been shown to produce an artifact of methyl oleate and the amount of artifact observed may increase with an increase in the "age" of the reagent and increase with reaction time (83). Methanolysis with sodium methoxide (84) was used as a standard method in the study of Fulk and Shorb (83).

Methylation of fatty acids with diazomethane (85) is another very popular, quick and mild procedure for insect lipids (35,74,86,87). Chloro- form:methanolic HCl:cupric acetate was used to esterify fatty acids at room temperature (88).

In an effort to overcome the problems of artifacts, and loss during boiling and evaporation, a number of transesterification or more specifically transmethylation procedures have been developed and used on insect lipids. A very mild technique for transesterification of esters of glycerol, be they glycerides or phospholipids, is accomplished in chloroform:methanol:NaOH solution at room temperature according to the technique of Morgan, et al. (89) added to by Mares (90) and utilized on insects by Fast (4) and Jackson et al. (86). Renkonen (91) suggests using either NaOH or HCl in chloroform:methanol for the transesterification of phospholipids of insects (92). Lambremont (93) modified older concepts of using 2,2-dimethoxypropane to drive transesterification to completion at lower temperatures (94,95) and came up with refluxing up to 20 mg of lipid under N_2 in 20 ml anhydrous methanol containing 0.5 ml sulfuric acid and 1.0 ml of 2,2-dimethoxypropane for 4 hours at 70 C.

To avoid excess handling, loss on adsorbents and contamination from large volumes of solvents, Christie (96) developed a technique of esterification of triglyceride, diglyceride, monoglyceride, free fatty acid and phospholipid bands performed on the thin-layer adsorbent itself. Davison and Dutton (97), also in an effort to avoid multiple-handling techniques and consuming relatively large quantities of samples and time, developed a microreactor method for methanolysis. Although not widely used in insect lipid analysis in the past, these last techniques may be useful as researchers are investigating smaller amounts of lipid from insect tissues.

2) GLC of fatty acid methyl esters.

Following methylation, the method of choice to analyze fatty acids is GLC at least on a polyester phase column and sometimes on a silicone phase column. The two most popular polyester phases are ethylene glycol succinate(EGS) and diethylene glycol succinate (DEGS). The six foot 15% EGS column according to Horning, et al. (98) has been used for insect fatty acids (27) and shorter versions were used by Fast (4) and Sloan et al. (99). With the advent of better solid phases the use of a 6 foot 6% DEGS column (12,32,40,80,100) or 10 to 20% DEGS (13,101,102) has become popular. Lambremont (93) used a 550 cm 20% diethylene glycol adipate and 2% phosphoric acid column. The use of packed columns is not universal, Madariaga et al. (43) used a 50m x 0.25 mm capillary column of DEGS.

3) Identification of the fatty acid methyl ester peaks.

With some experience and proper selection of a reliable GLC system most of the fatty acid methyl ester peaks can be identified by comparison of their retention times with those of standards (4,12,13,27,32,40,80,99, 100,101,102). A second method of identification or characterization is to plot the logarithm of retention times vs carbon number (103,104). This technique is especially useful when using a silicone phase column (4,27, 87).

4) Quantitation of the fatty acid methyl esters.

Quantitation of the recorder signal from the gas chromatograph is generally obtained by determining the area under the chromatograph peak. This can be accomplished by: 1) Disc integration as used by Thomas (32), Hutchins and Martin(40), and Fast (4), 2) Electronic integration as used by Madariaga et al. (43), 3) Planimetry (27,105) and 4) Triangulation (13,40, 80,100,102). Multiplying the retention time of the peak by the peak height (106) will give an estimate of the peak area if the column has been operated isothermally (4). Beenakkers and Scheres (12) added known quantities of methyl heptadecanoate as an internal standard.

Various TLC techniques have been employed to analyze and to separate appreciable quantities of saturated, monoene, diene, triene, etc. fatty acid methyl esters. The separation of fatty alcohols from fatty acid methyl esters can be achieved on silica gel TLC plates developed in n-hexane: ether:glacial acetic acid (80:20:1 v/v/v) (93,107). The fatty acid methyl esters can be separated according to unsaturation on silica gel impregnated with $AgNO_3$ (3,108) or silica gel treated with ammoniacal silver ion followed by development in toluene:isooctane:n-hexane (80:10:10 v/v/v) (74,109). Visualization of the chromatogram is obtained by spraying the plate with sulfuric acid:dichromate and heating at 120-180C (60,74).

VI. STEROL ANALYSIS

It is generally agreed that insects lack the capacity for de novo biosynthesis of the steroid nucleus; therefore, the sterols isolated from insects consist of dietary sterols, sterols on the metabolic pathways to cholesterol, cholesterol, Δ^7-sterols, sterol intermediates of ecdysones, ecdysones and others.

One of the most referenced techniques for insect sterol analysis is the procedure outlined by Svoboda, Thompson and Robbins (110). The procedure suggests sacrificing the insects by freezing, then extracting the lipids with chloroform:methanol (2:1 v/v) using a Virtis homogenizer or a Ten Broeck tissue grinder (111). After the extracts are filtered and partitioned with water, the chloroform phase is taken to dryness on a rotary evaporator. The partitioning is similar to the Folch technique (2). The extract is fractionated on a silicic acid column using increasing benzene in hexane followed by chloroform and methanol. They suggest 20 mg lipid/gram of column adsorbent. An alternative column chromatography would be that of Hirsch and Ahrens (56). The column fractions are monitored by TLC on silicic acid developed in benzene mixed with ethyl acetate, ethanol, or cyclohexane (63,112). The sterol esters elute from the silicic acid column in 20% benzene in hexane and the free sterols elute in the benzene and chloroform fractions. Rechromatography of the sterols (or saponified sterol esters) on neutral grade IV alumina (113) or neutral grade II alumina (114) eluting with increasing benzene in hexane provides a preparation suitable for GLC.

When only total sterols are of interest, it is possible to: 1) Homogenize the insects, 2) Folch extract the homogenate and 3) Saponify the extract (115). Alternatively the insects can be homogenized and the total homogenate saponified (116). The non-saponifiable lipid can be obtained from the saponification mixture by removing the solvent in vacuo, and extracting the residue with ether. The non-saponifiable lipids are fractionated into hydrocarbons, alcohols and sterols by chromatography on alumina (117) or on silicic acid (118). Martin and Carls (119) used an ion exchange resin eluted with wet ether to free the sterols of fatty acid contaminants which persisted through the purification procedure. The sterol fraction can then be acetylated (112,120) or taken in the free form to GLC.

The preparation of sterol-digitonides is a useful method to remove non-steroid components from sterol containing column fractions or preparative TLC fractions. Various modifications of the methods of Moore and Baumann (121) or Sperry and Webb (122) are used for preparing sterol digitonides. Schaefer, et al. (27) modified the Sperry and Webb technique to include careful washing of the digitonides with 85% ethanol, ethanol:acetone (1:1), acetone:ether (1:1), and ether. The digitonides are split by heating in pyridine (123) and extracted in hexane.

Recent studies of the Mexican bean beetle, Epilachna varivetis, have shown that sterols other than cholesterol have a significant role in sterol metabolism in this specie (111). The presence of lathosterol (Δ^7-cholesten - 3β-ol) comprised 12-16% of the sterols from eggs, pre-pupae and adults; and Δ^7-campestenol and Δ^7-stigmastenol were also identified in all stages. Surprisingly cholesterol accounted for only 2.9 to 4.5% of the total sterols in the stages examined. Findings such as these should prompt researchers to more thoroughly characterize the sterols isolated from insects. The commonly used GLC and TLC systems do not always readily resolve a number of the saturated and unsaturated sterols.

Techniques useful in resolving saturated and unsaturated sterols include TLC and column chromatography using $AgNO_3$ impregnated silica gel
or alumina and GLC on both a silicone phase and a polyester phase.
Cholesterol propionate was separated from desmosterol propionate and
fucosterol propionate on chromatoplates of 25% $AgNO_3$ impregnated alumina
developed in hexane: ethyl acetate (25:1 v/v) (124). Goodnight and Kircher
(115) monitored column eluants of free sterols on TLC plates of 10% $AgNO_3$
on silica gel developed in chloroform:acetone (95:5 v/v). Svoboda, et al.
(111,125,126) used a column of 20% $AgNO_3$ impregnated silicic acid eluted
with increasing benzene in hexane and monitored the column eluants on a
20% $AgNO_3$ impregnated silica gel TLC system to separate sterol acetates.
In this system they obtained three distinct bands consisting of saturated
sterol acetates, monounsaturated sterol acetates and diunsaturated sterol
acetates. The Δ^5-sterol acetates were only partially separated from lathosterol acetate (Δ^7 - sterol acetates) using this system.

GLC of the sterols on a silicone column or GLC of the sterol acetates,
propionates or trimethylsilyl ethers (119) on a silicone column and a polyester column is useful to detect the presence of certain sterols. Gas chromatography - mass spectrometry (GC-MS) substantiate the presumed identity
of a sterol within a GLC peak. Table 1 contains relative retention times
for a number of sterols and sterol acetates on several GLC columns. A
recent extensive evaluation of fourteen GLC liquid phases used twelve
acetylated plant sterols many of which have been observed in insects
(127). The phases were rated according to their ability to resolve pairs
of sterols regarded as difficult to separate.

Although the bulk of the insect sterol analyses has involved whole
insects, a few studies were concerned with intracellular distribution of
sterols (128,129). The researchers incorporated radiolabeled sterols
into the diet of the insects, then separated the various tissues into
nuclear, mitochondrial and microsomal fractions by differential centrifugation. The radioactive sterols were then extracted, purified and counted
to determine total incorporation (128). The sterols were then chromatographed on alumina and GLC, and the GLC effluent was collected and
counted by liquid scintillation. Minimal redistribution of labeled sterols
incorporated under long-term in vivo conditions occurred during the usual
homogenization and fractionation procedure (129).

TABLE 1: Relative GLC Retention Times for Several Sterols and Sterol
          Acetates Relative to Cholestane

	A	B	C	D	E	F
Cholestane	1.00	1.00	1.00	1.00	1.00	1.00
Cholesterol	1.83	1.83	1.87	2.81		
β-Sitosterol	2.91	2.91	3.06	4.34		
Fucosterol			3.03	4.56		
Cholesterol acetate	2.57	2.58	2.76	4.64	3.41	6.63
Cholestanol acetate					3.41	6.51
Lathosterol acetate					4.00	8.01
7-Dehydrocholesterol acetate	2.81					
Campesterol acetate			3.62	6.01	4.51	8.88
Campestanol acetate					4.50	8.78
Fucosterol acetate			4.33	7.32		
Stigmasterol Acetate					4.90	9.42
Δ^7Campestenol acetate					5.20	10.70
β-Sitosterol acetate	4.18	4.15	4.53	7.20	5.63	11.14
7-Dehydro-β-sitosterol acetate	4.80					
Stigmastanol acetate					5.61	11.02
Δ^7-Stigmastenol acetate					6.45	13.50
24-Methylene cholesterol			2.31	3.50		
24-Methylene cholesterol acetate			3.24	5.65		

Column A.   3.5ft x 0.10 in (ID) column of 3.0% SE-30 operated at
            235C (113).

Column B.   6ft x 0.19 in (ID) column of 1.5% SE-30 operated at 235C
            (117).

Column C.   6ft x 0.10 in (ID) column of 0.75% SE-30 operated at 230C
            (110,133).

Column D.   6ft x 0.19 in (ID) column of 3.0% QF-1 operated at 230C
            (110,133).

Column E.   6ft x 0.19 in (ID) column of 1.0% OV-17 operated at 230C
            (111,113).

Column F.   6ft x 0.10 in (ID) column of 0.75% Neopentyl Glycol Succi-
            nate operated at 215C (110,111).

Sterol esters have been observed in measurable quantities in a number of insects. Sterol esters accounted for about 41 percent of the total sterol of housefly eggs (130). Sulfate esters of cholesterol, campesterol and β-sitosterol were identified from the meconium of the tobacco hornworm, Manduca sexta (131). Sterol sulfates have a strong band in the infrared spectrum at 1200 cm^{-1} and have a distinctive mass spectra (131).

Beyond the scope of this chapter is the analysis of ecdysones derived from insect sterols. A recent review on the metabolism and utilization of phytosterols and other neutral sterols, the in vivo metabolism of the ecdysones, and the in vitro metabolism of ecdysones cites many references that would be useful in ecdysone analysis (132).

VII. POLAR LIPID ANALYSIS

The polar lipids are quantitatively separated from the neutral lipids on silicic acid column chromatography by first eluting with chloroform or a less polar solvent to remove the neutral lipids, then eluting with methanol or a mixture of methanol:chloroform to elute the polar lipids (50). The major phospholipids of insects are phosphatidyl ethanolamines (cephalins) and phosphatidyl cholines (lecithins) (4). Phospholipids present in lesser amounts or in a limited number of insects include sphingomyelins, phosphatidyl serines, lysophosphatidyl cholines, phosphatidyl inositols, phosphatidic acids, cerebrosides, and others (134,135,136,137,138,139).

This wide variety of polar lipids, many with quite similar polarities, has necessitated the development of numerous column and thin layer chromatographic systems to provide complete separation of these lipids. Several papers concerned with the analysis of phospholipids are frequently referenced in papers dealing with insect polar lipids. Skipski et al. (140) used one-dimensional TLC with chloroform:methanol:acetic acid:water (25:15:4:2 v/v) as the solvent system. They were able to separate and quantitate lysophosphatidyl cholines, sphingomyelins, phosphatidyl cholines, phosphatidyl inositols, phosphatidyl serines, phosphatidyl ethanolamines and a less polar fraction which included cardiolipins. A two-dimensional system developed by Abramson and Blecher (141) separates all the polar lipids in the study by Skipski et al. (140) plus additional ones lacking nitrogen. They used chloroform:methanol:acetic acid:water (250:74:19:3 v/v) as the first solvent, then chloroform:methanol: 7$\underline{M}$ NH$_4$OH (230:90:15 v/v) as the second solvent. Rouser and coworkers have published several papers describing the analysis of polar lipids (142,143,144,145).

These systems do not separate polar lipids containing ether linkages from polar lipids containing only ester linkages. Lambremont and Wood (139) used the methods of Wood and Snyder (55) to identify glyceryl ether phospholipids in several insects. Fast (4) was primarily concerned with the principle phospholipids, phosphatidyl ethanolamines and phosphatidyl cholines. Consequently, he used a diethylaminoethyl (DEAE)-cellulose column to separate these lipids. On the other hand the total analysis of phospholipids from some species has required two-dimensional TLC (102, 146) or development on two different TLC plates using two different solvent systems (36). Two-dimensional TLC is more time consuming and more expensive than one-dimensional TLC. It is advantageous to do the preliminary analyses using two-dimensional TLC to ascertain the number of components and their mobilities; but if the simplicity of the mixture allows, a one-dimensional TLC system should then be adopted. Table 2 lists TLC systems used in some of the analyses of insect and other animal polar lipids.

Polar lipids are identified on the basis of their chromatographic properties and their reactivity to various spray reagents. Co-chromatography with authentic standards of phospholipids and ratios of phosphorus to fatty acids are also used for identification. Various spray reagents used to characterize the polar lipid groups include: Ninhydrin reagent, Rhodamine 6-G, 2,5-dichlorofluorescein, Dragendorf reagent, phosphomolybdic acid, anthrone reagent, iodine vapor and sulfuric acid (with and without dichromate). The ninhydrin spray solution helped to identify nitrogen-containing polar lipids (147). The phosphomolybdic acid spray is to confirm the presence of phosphorus (147). Dragendorf reagent confirms the presence of choline in phosphatidyl cholines, sphingomyelins and lysophosphatidyl cholines (147). Anthrone reagent is used to detect cerebrosides (148). Iodine vapor, Rhodamine 6-G, and 2,4-dichlorofluorescein are non-destructive and are not specific for polar lipids. The spots can be marked with a needle to provide a more permanent record. Plates sprayed with sulfuric acid:dichromate when charred, develop black spots where the lipids are present. These spots can be quantitated using photodensitometry (62,149,150). The charred spot can also be scraped from the TLC plate and analyzed for inorganic phosphorus (145). Unreacted scrapings from TLC plates are directly analyzed for phospholipid phosphorus (151,152,153,154).

TABLE 2:   Examples of polar lipid separations obtained using thin layer
           chromatography

Polar Lipids		Solvent System	Reference
		1-Dimensional	
PC,PE,Sph,LPC,		$CHCl_3:CH_3OH:H_2O$	(137)
Cer,PI,PA		(65:25:4 v/v)	
PGP,PE,PS,PC		$CHCl_3:CH_3OH:7\underline{N}\ NH_4OH$	(155)
Sph,PI		(115:45:7.5 v/v)	
PA,Cer,		$CHCl_3:CH_3OH:7\underline{N}\ NH_4OH$	(138)
PE,PC,		(60:35:5 v/v)	
PS,Sph,	or	$CHCl_3:CH_3OH:H_2O$	
LPE,LPC		(84:32:3 v/v)	
PE,PS,PC,		$CHCl_3:CH_3OH:H_2O:CH_3COOH$	(156)
Sph,LPC		(65:25:3:4 v/v)	
LPC,Cer,PC,		$CHCl_3:CH_3OH:CH_3COOH:H_2O$	(157)
PS,PI,PE		(25:15:4:2 v/v)	
		2-Dimensional	
PC,PE,		$CHCl_3:CH_3OH:7\underline{N}\ NH_4OH$	(146)
Sph,		(65:30:5 v/v)	
LPC,	then	$CHCl_3:CH_3COCH_3:CH_3OH:CH_3COOH:H_2O$	
PI		(50:20:10:10:5 v/v)	
LPC,Sph,		$CHCl_3:CH_3OH:7\underline{N}\ NH_4OH$	(102)
PI,PS,		(60:30:5 v/v)	
PC,PE,	then	$CHCl_3:CH_3OH:CH_3COOH:H_2O$	
PGP		(65:25:8:3.5 v/v)	

Abbreviations: PE-phosphatidyl ethanolamines, PC-phosphatidyl cholines,
Sph-sphingomyelins, PS-phosphatidyl serines, LPC-lysophosphatidyl
cholines, LPE-lysophosphatidyl ethanolamines, Cer-cerebrosides, PA-phos-
phatidic acids, CL-cardiolipins, PI-phosphatidyl inositols, PGP-poly-
glyceryl phosphatides.

VIII. LIPOPROTEINS OF INSECT HEMOLYMPH

In experiments with <u>Hyalophora cercropia</u>, it was observed by Chino and Gilbert (158) that diglycerides released from the fat body into the hemolymph were associated with specific hemolymph lipoproteins involved in the release and transport phenomena. The hemolymph lipoproteins from several species of Lepidoptera have been purified and analyzed (159,160, 161,162,163). The hemolymph lipoproteins were labeled with [^{14}C]-palmitate as a radiotracer during the purification and characterization procedure (159,160). Chino, <u>et al</u>. (160) used ammonium sulfate fractionation, precipitation at low ionic strength (pH 6.5) and DEAE- cellulose column chromatography to purify the labeled lipoproteins. Thomas and Gilbert (161) used ultracentrifugal fractionation with solutions of varying density to isolate and purify lipoproteins, and determined flotation and sedimentation constants of the lipoproteins by analytical ultracentrifugal analysis. By sedimentation equilibrium analysis Chino, <u>et al</u>. (160) determined the molecular weight of two of the principle diglyceride carrying lipoproteins to be approximately 700,000 daltons for lipoprotein-I (LP-I) and 500,000 daltons for lipoprotein-II (LP-II). Electron microscopic studies indicates that the two proteins are globular with calculated diameters of 13.5 ± 0.6 nm for LP - I and 10.0 ± 0.5 nm for LP - II, and may have substructure (160).

Lipids constitute 44% of the weight of LP-I and 10% for LP-II (160) as determined by Folch extraction (2). The lipids were fractionated by column chromatography, and TLC analysis showed that most of the lipids are diglycerides, phospholipids and sterols (160,161).

Electrophoresis of hemolymph from seven species of Lepidoptera showed that all species examined have hemolymph lipoproteins, and in five of the seven species the lipoproteins are also glycoproteins (163). The electrophoresis was done on polyacrylamide using a high resolution slab gel apparatus. The slab gel technique was used instead of disc gel electrophoresis so that exact comparisons between the proteins could be made. The lipoproteins were stained with lipid crimson and destained with acetic acid. The glycoprotein stain was a modified periodic acid - Schiff procedure (164).

IX. INSECT EGGS

There have been relatively few studies on the lipid composition of insect eggs or embryos; however, with increasing interest in the role of lipids in insect vitellogenesis, the analysis of insect egg lipids may occur more frequently.

The analysis scheme is generally similar to that used for total lipid or fat body lipid analysis. Homogenization in a glass tissue grinder with chloroform:methanol (2:1 v/v), followed by filteration, and the Folch et al. (2) wash are commonly used (86,130,165,166,167,168). Following evaporation of the solvent and total lipid determination by gravimetric means, the lipids are usually taken to column chromatography. Sroka and Barth (165) used a technique derived from Barron and Hanahan (169), while others use the method of Horning (118). Suzuki et al. (166) used increasing benzene in hexane followed by increasing ether in hexane to elute the lipids. The column fractions were monitored by TLC on silica gel using petroleum ether:ether:acetic acid (90:10:1 v/v).

Fatty acid methyl esters are usually thought of as derivatives of fatty acids prepared for GLC analysis; however, fatty acid methyl esters have been observed from a number of insect eggs (86,99,166). McFarlane and others (reviewed by McFarlane 170) have shown that fatty acid methyl esters play a role in control of growth and development. Since it is possible to carry out transesterification in either acid or alkaline chloroform:methanol (see the section on methylation), it is important to take precautions to prevent the formation of artifacts. Sloan et al. (99) used chloroform:ethanol and chloroform:isopropanol (2:1, v/v) instead of chloroform:methanol for extraction.

The organic solvents used for both extraction and chromatography were treated to remove any contaminating low molecular weight alcohols (171) and were tested by GLC for contaminating methanol. Jackson et al. (86) showed that the fatty acid methyl esters could be extracted with methanol free hexane, or with chloroform:ethanol (2:1 v/v). Suzuki et al. (166) checked their chloroform: methanol extraction procedure by adding a radiotracer of [1-^{14}C]-palmitic acid. They found that only a trace of [1-^{14}C] -- palmitic acid was incorporated into fatty acid esters during their procedure. It was concluded that fatty acids were not esterified by refluxing with solvents such as methanol, chloroform:methanol (2:1 v/v), methanol:water (95:5 v/v), chloroform:ethanol (2:1 v/v) and ethanol. Likewise fatty acid methyl esters were not transesterified by refluxing with chloroform:ethanol (2:1 v/v), ethanol, chloroform:isopropanol (Suzuki et al. 1970).

X. CUTICULAR LIPIDS

Those lipids which can be extracted from the surface of intact insects
or from exuviae are cuticular lipids. There is general agreement that their
primary function is to form a water impermeable barrier on the surface of
the cuticle. They are, therefore, an adaptation to the terrestrial environ-
ment. Other functions are known or suspected, including: 1) A communicative role, as semiochemicals (172), 2) Mechanical protection from abra-
sion, and 3) As a physical barrier to some microorganisms. In addition
the cuticular lipids affect the penetration of insecticides into the insect,
although whether the affect is positive or negative remains debated (173,
174).

Generally speaking the cuticular lipids of an insect may include hydrocarbons, free fatty acids, free alcohols, sterols, wax and sterol esters,
and triacylglycerols. In addition cuticular lipids may also contain traces
of other lipids such as pigments and polar lipids (175,176).

The extraction of cuticular lipid has been accomplished under various
conditions. Recent procedures include extraction in hexane for ten minutes
(177), for fifteen minutes (178,179), and for unspecified time using sintered glass funnels (180,181,182). Jackson (180,181) followed these hexane
extractions with similar chloroform extractions to assure complete extraction. Silhacek et al. (183) poured diethyl ether over the insects for a very
brief extraction period. Goodrich (184) extracted with methylene chloride,
first for three minutes, then overnight. Nelson and Sukkestad (185) homogenized exuviae in chloroform:methanol (2:1 v/v). Gilby and McKellar (186)
extracted eight times with chloroform, then refluxed the puparia for fifteen
minutes in chloroform:methanol (1:1 v/v). Thompson and Barlow (101) extracted dissected cuticles by the procedure of Bligh and Dyer (34). Jackson,
Armold and Regnier have used a KOH digestion-hexane extraction procedure
in studies on cuticular hydrocarbons of Sarcophaga bullata (105,182,187).

With intact insects one is confronted with two conflicting realities:
1) Short extraction periods may not remove the cuticular lipid quantitatively, and 2) Long extraction periods may extract internal lipids. One can
extract the intact insects for varying periods of time and analyze the extracts by TLC to determine when a qualitative change occurs in the extracts
(178,188). Even so, selection of the longest time period from which the
extract lacks internal lipid does not guarantee that all of the cuticular lipid
has been extracted. With exuviae it is generally assumed that all of the
extracted lipid is cuticular in origin; however, extracts of excised cuticle

may contain lipid from other tissues transferred during removal.

The separation of cuticular lipids is similar to the separation of other neutral lipids. Recent use of adsorption column chromatography using silicic acid include Gilby and McKellar (186), Goodrich (184), Jackson (180, 181), Blomquist et al. (178), Silhacek et al. (183), and Armold and Regnier (105,187). In addition, Soliday et al. (179) and Jackson et al. (182) used silicic acid column chromatography followed by preparative TLC. Nelson and Sukkestad (185) used Florisil column chromatography to purify the alkanes of exuviae from Schistocerca vaga. Armold et al. (188) and Lok et al. (177) used preparative TLC to purify the cuticular lipids of the big stonefly and two fire ant species.

Hydrocarbons are often an abundant component of cuticular lipid. They have also been reported from hemolymph (189,190), fat body (190,191), whole body extracts (192, 193), and specific tissues or parts of insects (46). In addition, several hydrocarbons or mixtures of hydrocarbons have pheromonal functions and are discussed in another chapter.

Once the hydrocarbons have been purified by column chromatography and/or preparative TLC, the purity of the fraction should be determined by analytical TLC (177). The purified hydrocarbon fraction can then by analyzed for the presence of unsaturated hydrocarbons by TLC using silver nitrate impregnated plates (177,179), by infrared spectroscopy (184,190), by GLC analysis using bromination to remove unsaturated hydrocarbons (185,188, 190), and by GLC analysis using hydrogenation to alter the retention times of any unsaturated hydrocarbons (46). If unsaturated hydrocarbons are present, they can be separated using silver nitrate impregnated silicic acid column chromatography (192,193) or preparative TLC using silver nitrate impregnated silica gel plates (180,181,182,188).

Structural analysis of unsaturated hydrocarbons involves a variety of techniques. Hydrogenation followed by GLC analysis was used to determine chain lengths by Hutchins and Martin (192). The stereochemistry of the double bonds can be determined by infrared spectroscopy (180,194), and NMR spectroscopy was used in the assignment of the stereochemistry about the double bonds of the alkadiene from the American cockroach (194, 195). To determine the position of double bonds, Jackson (180) used osmium tetroxide oxidation followed by periodate-permanganate oxidation to cleave at the double bond(s). The resulting acids were methylated (85) yielding fatty acid methyl esters which were then analyzed by GLC. Various additional periodate-permanganate oxidations have been used (192,194,195). A

mass spectral analysis of the unsaturated hydrocarbons of the Japanese
and American cockroaches was accomplished by first forming the diols with
osmium tetroxide oxidation, followed by formation of the trimethylsilyl
ethers with bis-(trimethylsilyltrifluoracetamide (181). The ethers were
then analyzed by MS.

The alkanes are usually analyzed by GLC and MS. The n-alkanes are
identified as those peaks which either co-chromatograph with authentic
standards or whose retention times fall on a graph plotting retention times
versus number of carbons in n-alkane standards (46,179,184,185,192,193).
They are also identified as n-alkanes if they are removed by 5A molecular
sieve (46,179,185,192,193) or by urea inclusion (184). Martin and Mac-
Connell(193) identified n-nonacosane by comparison to its published mass
spectra. Those alkanes which are not removed by 5A molecular sieve and
do not co-chromatograph with n-alkane standards are branched alkanes.
Although partial and tentative characterization can be made by comparing
the retention times to those of 2-methyl, 3-methyl and internally branched
monomethyl alkanes (177,190), further characterization requires the application of mass spectrometry. Armold et al.(188), Jackson (181),Jackson et
al. (182) and Lok et al. (177) used combined gas-liquid chromatography-
mass spectrometry systems. Jackson (180), Cavill et al.(46), Soliday et
al. (179) and Nelson and Sukkestad (185) used preparative GLC to obtain
the samples that were subsequently analyzed by MS. Since the interpretation of these mass spectra is, at times, involved, readers are urged to
read the original papers. New mass spectral procedures have been applied
to insect hydrocarbon analysis; Jackson et al. (182) used mass chromatographic techniques to identify numerous monomethyl isomers from S. bullata.
Multiple mass spectra were obtained from several different points within
any given GLC peak. Since the GLC retention times of the isomers vary
slightly, the mass spectra are not identical. Each mass spectrum has its
own ion maxima characteristic of the isomers predominating within that
portion of the peak.

The analyses of cuticular free fatty acids and cuticular triacylglycerols (177,179,182,184,186,188) are similar or identical to the procedures discussed in the Fatty Acid Analysis section. Similarly, cuticular
sterol analysis parallels those procedures covered in the Sterol Analysis
section. Long chain fatty alcohols have been reported from the eri silk-
worm (196). Infrared spectroscopy, melting point determination, elemental
analysis, chemical derivations and GLC and TLC analyses were used to

characterize these lipids. Wax and sterol esters have been reported from
several insects (177,178,182,184,186,188). The wax esters of Melano-
plus sanguinipes and M. packardii are esters of secondary alcohols as
determined by chemical reductions, GLC analysis and mass spectrometry
(178). Lipids from the surface of scale insects and bees are considered
as cuticular lipids in some investigations, and these are discussed in the
next two sections of this chapter.

XI. SCALE INSECT LIPIDS

Some members of the scale insect group are useful to man, yielding
such important commercial products as dyes, varnishes (shellac) and
waxes. Others, due in part to their reproductive potential, genetic adapt-
ability, and feeding habits are classified as destructive agricultural pests.
These insects are covered with a waxy plaque or scale from which the
common name scale insect is derived. For a review on the chemistry of
scale insects and aphids see Brown (197).

Most of the recent work on scale insects has been done on the genus
Ceroplastes, although earlier investigations were carried out on the wax
of Dacylopius coccus (198), Laccifer lacca (199), Gascardia madagascaren-
sis,Coccus ceriferus and Tachardis lacca (200). Faurot-Bouchet and
Michel (201) found a surprisingly large percentage of 26-34 carbon fatty
acids which required special attention when separating acidic and neutral
substances since the salts of these very long chain fatty acids are not
appreciably soluble in water. They used the procedure of Savidan (202)
which involves: 1) Dissolving the saponification products in warm chloroform or warm benzene, 2) Passing the solution through a column of sodium
carbonate to bind the salts while the neutral substances elute, and 3)
Dissolving the sodium carbonate in acidified water from which the precipitated fatty acids could be separated by filtration. The dicarboxylic acid
tetradecan-1,14-dioic acid from the Comstock mealybug, Pseudococcus
comstocki, was purified by chromatography of the methylated fatty acids
on silicic acid eluted with n-hexane:diethyl ether (94:6 v/v) (203). The
acetone recrystalized crystals were characterized by melting point, carbon-
hydrogen analysis, urea adduct formation and other chemical methods. The
infrared spectrum, melting point and chromatographic characteristics were
compared to a synthetic preparation of methyl tetradecan-1,14-dioate.

Early work on Ceroplastes was carried out by Kono and Maruyama (204),
Hackman (205), Gilby and Alexander (206) and Gilby (207). Since 1960

most of the investigations of Ceroplastes waxes have been reported from
the laboratories of T. Rios, et al., Y. Tamaki, et al., and Hashimoto, et al.
From Ceroplastes albolineatus, Rios and Colunga (208) isolated three
new alcohols which they named ceroplastol I, ceroplastol II and albolineol.
To isolate the alcohols, the crude wax was saponified and separated into
acidic and neutral components. The neutral components were chromatographed to yield fractions which were characterized by melting point,
melting point of the 3,5-dinitrobenzoate derivative, optical rotation, ultraviolet absorption, and NMR spectra. Further NMR, mass spectra and x-ray
crystal structure data on ceroplastol I and ceroplasteric acid were collected
by Iitaka et al. (209). Ceroplastol II was further characterized by NMR,
infrared spectroscopy and isomerization to ceroplastol I by Rios and Quijano (210). Albolic acid, also isolated from Ceroplastes albolineatus,
was characterized by its optical rotation, infrared absorption, ultraviolet
absorption, and NMR spectrum as well as isomerization of ceroplasteric
acid to albolic acid (211). Albolineol was found to be a bicyclic sesterter-
pene as characterized by melting point, optical rotation, ultraviolet and
infrared absorption, NMR spectra and mass spectra of the original compound as well as a number of derivatives (212). The acyclic C_{25} isopre-
noid alcohol geranylfarnesol, which is proposed as a biogenic precursor of
the sesterterpenes, has been isolated and characterized from C. albolineatus (213). Geranylfarnesol was characterized by identifying the aldehydes
resulting from chromium trioxide-pyridine oxidation and previously used
techniques (213).

Tamaki et al. used chemical means, x-ray diffraction (214) and electron microscopy (EM) (215) to investigate a number of scale insect waxes,
in particular C. pseudoceriferus (216), C. japonicus, C. rubens (217,218)
and Pseudococcus comstocki (203). The chemical techniques used were:
1) Extraction with an organic solvent (usually chloroform), 2) Precipitation of the "hard wax" with ethanol, leaving a "soft wax" fraction, 3)
Saponification of the wax fractions, 4) Formation of urea adducts of the
straight chain alcohols, and 5) Direct GLC of the fatty alcohols and GLC
of the fatty acids as methyl esters. The x-ray diffraction studies were performed on chloroform extracted wax which was then dried and analyzed by
a routine x-ray powder technique (214). Scanning electron microscope observations were made in situ and of external dermal structure (215). Transverse sections of the insect body were studied by EM. For the external
observations of dermal structures by scanning EM, the samples were fixed

with 10 percent formalin, dehydrated with an n-butanol:ethanol system,
and thoroughly washed with xylene. Transverse sections of samples were
fixed in 10 percent formalin or 0.5 percent osmium tetroxide solution,
washed with water, dehydrated with n-butanol:ethanol, and embedded in
melted paraffin. The paraffin sections of 5-10μ thickness were stained
with Hansen's Haematoxylin-eosin. Frozen sections of intact insects
were stained with Nile blue sulfate or Sudan black B.

The more recent Hashimoto et al. papers on the lipids of scale insects
report the triacylglycerol fatty acid distribution (219,220) and the hydrocarbon, free fatty acid and wax ester composition (221,222,223). The
triacylglycerols were extracted according to the Folch et al. (2) technique,
chromatographed on silicic acid, and hydrolysed with pancreatic lipase.
The hydrolysis products were chromatographed by TLC and followed by GLC.
The triacylglycerol types (SSS,SSU, etc.) were calculated according to 1,
3-random-2-random distribution, ordered distribution and 1,2,3-random
distribution theories. The hydrocarbons, wax esters and free fatty acids of
Ceroplastes rubens, C. japonicus and C. pseudoceriferus (221) were separated by silicic acid chromatography. The saturated and unsaturated components were separated on silicic acid impregnated with silver nitrate.
Following methylation, the fatty acids were characterized by GLC on DEGS,
while the hydrocarbons were chromatographed directly on SE-30 and OV-1
GLC columns (222,223).

The Indian scale insect Laccifer lacca produces a secretion which
accumulates along infested branches, eventually surrounding the insects
and ultimately leaving a thick cylindrical residue called lac. Although a
number of varieties of lac exist, depending upon the host plant, all are
physically and chemically very similar. Industrial processing yields the
shellac of commerce.

The principle known aliphatic acids of lac resin are threo-aleuritic
acid (9,10,16-trihydroxyhexadecanoic acid) (224,225), and butolic acid
(6-hydroxytetradecanoic acid) (226,227). Minor acids detected include
myristic, palmitic, palmitoleic, and other C_{14-18} straight chain acids,
C_{16} ω-hydroxy acids, C_{14} and C_{16} 9,10-dihydroxy acids, and 6-oxotetra-
decanoic acid (224,226).

Lac resins are complex oligomers of the above hydroxy-acids and a
series of sesquiterpenic aldehyde-acids joined by ether (228) and ester
linkages. The principle sesquiterpenic aldehyde-acids are jalaric acid
(225,229) and laccijalaric acid (230). However when the resin is

hydrolyzed by alkali, especially at elevated temperature, the possibility
of a Cannizzaro type disproportionation to a whole series of other acids is
possible (224,225,229,230,231,232,233). There is some question whether the other acids (e.g. epishelloic acid) are truly artifacts since they
have been detected in untreated lac resin (234). The analysis and structure elucidation of the above mentioned acids is extensive and will not be
illuciated here. The reader is referred to the papers cited for that technology. The modern TLC techniques are invaluable for separation of lac acids
but GLC has proven to be of limited utility since the methyl esters of many
of the lac acids decompose on the GLC columns (233). The use of the
recently developed ambient temperature high pressure liquid chromatographic techniques could contribute significantly to future studies of lac
resins.

Hard and soft lac resins appear to contain the same acids, but have
varying proportions of the acids and different intermolecular bonds which
alter the physical properties. However, when the hard resin was fractionated by precipitation from benzene: dioxane mixtures, twelve fractions of
quite different content and proportion of the constituent lac acids was
observed (235). Further purification of one fraction by progressive solution in ethyl acetate: dioxane and precipitation from dioxane with benzene
gave an oligomer that was submitted to further structural studies (236).
The reader is referred to the original papers for the techniques used to
characterize the oligomer. The results indicate that the oligomer has a
molecular weight of 2432, comprising two molecules of aleuritic acid,
one linked through the carboxyl (with all three hydroxyls free), and one
probably linked through the 9- and 16-hydroxyls; three molecules of laccijalaric acid, one linked through the 5-hydroxyl and possibly the carboxyl,
one linked through the carboxyl only, and one linked through the 5-hydroxyl
and the $-CH_2OH$ group, and possibly the carboxyl; in all, eight intermolecular bonds are present (235,236).

XII. BEESWAX

The analysis of beeswax has received limited attention, with the
primary thrust of these investigations directed towards determining beeswax analytical values and composition. These studies have contributed
to the development of procedures to investigate whether paraffin, stearic
acid or other adulterants could be detected. Methods for saponification
number, acid number, ester number, color, melting point, flash point,
and saponification cloud test are reported in papers by White et al. (237)

and Tulloch and Hoffman (238). Many of these techniques are U.S.P.
methods (239) or methods specified by the American Wax Importers
and Refiners Association Inc., 225 West 34th Street, New York, N.Y. 1001.

Hydrocarbon quantitation is accomplished by chromatography on
alumina (240) or on silicic acid (238) followed by weighing the solvent
free residue. Hydrocarbon freezing points were determined (237) as described by White et al. (240). GLC of the hydrocarbons was carried out on
a 3 ft x 1/8 in stainless steel column packed with 2% SE-30 and operated
at 125 to 350C at 3 C/min (238). Odd numbered chain length hydrocarbons
predominate and even numbered chain length hydrocarbons are at a very
low level allowing for detection and estimation of adulterants (238).

Most reference books describe the esters of beeswax as being mainly
myricyl palmitate (triacontanyl hexadecanoate), but this is misleading
since Holloway (241) indicates that by GLC of the wax esters the principle
alkyl esters are $C_{40}, C_{42}, C_{44}, C_{46}$, and C_{48}. Tulloch and Hoffman (238)
found some C_{50} and C_{52} wax esters with the C_{46} and C_{48} wax esters
predominating. Hydrolysis (242) of the alkyl esters followed by GLC of
the products revealed that the esterified fatty acids differ from the free
fatty acids (238,241). The techniques for fatty acid and alcohol analysis
are outlined earlier in this chapter. Fractionation of beeswax and hydrolyzed beeswax by column chromatography and TLC reveals the presence of
diols and hydroxy acids in addition to the above mentioned components.
Tulloch (243) suggests that honey-comb cappings should be used since
bleaching and refining may alter the composition. Identification of many
of these fractions involves NMR spectroscopy (244), GLC, and examination
of their hydrolysis products (reviewed by Tulloch, 243).

XIII. CONCLUSION

The past fifteen to twenty years have brought great changes in the
technology of lipid analysis. The use of TLC and GLC has greatly increased the reliability and quantity of the results that the lipid investigator can obtain in a day's work. In this chapter we have not attempted to
cover the early development of insect lipid analytical technology, but
have concentrated our discussion to the methodologies reported in the
last ten years.

It can be concluded that the area of insect lipid analysis is a dynamic
one. Older techniques are being modified. New techniques are being
developed. Improvements in instrumentation are being made and new

instruments are coming into use.

Even though some advantages in comparison would be realized if all investigators used the same techniques, such a practice would stifle creative efforts which may ultimately develop and incorporate new technologies into lipid analysis.

Many of the readers of this chapter will likely conclude, as we have, that there isn't a single procedure of choice for many insect lipid analyses. Indeed, the purpose of this chapter is to facilitate the researcher's selection of those techniques which best suit their particular problem.

XIV. REFERENCES

1. D. H. Hunneman, Tetrahedron Letters 14(1968)1743.
2. J. Folch, M. Lees and G. H. Sloan-Stanley, J. Biol. Chem. 226(1957) 497-509.
3. H. E. Vroman, J. N. Kaplanis and W. E. Robbins, J. Insect Physiol. 11(1965)897-904.
4. P. G. Fast, Lipids 1(1966)209-215.
5. J. E. Kinsella and T. Smyth, Comp. Biochem. Physiol. 17(1966)237-244.
6. L. T. Wimer and R. H. Lumb, J. Insect Physiol. 13(1967)889-898.
7. A. M. T. Beenakkers and L. I. Gilbert, J. Insect Physiol. 14(1968) 481-494.
8. S. H. W. Cmelik, J. Insect Physiol. 15(1969)1481-1487.
9. R. D. Harlow, R. H. Lumb and R. Wood, Comp. Biochem. Physiol. 30(1969)761-769.
10. J. E. Kinsella, Lipids 4(1969)299-300.
11. R. G. Bridges and G. M. Price, Comp. Biochem. Physiol. 34(1970) 47-60.
12. A. M. T. Beenakkers and J. M. J. C. Scheres, Insect Biochem. 1(1971)125-129.
13. R. J. Cenedella, Insect Biochem. 1(1971)244-247.
14. G. M. Chippendale, Insect Biochem. 1(1971)39-46.
15. E. Cohen, Insect Biochem. 2(1972)161-166.
16. A. B. Dutkowski and B. Ziajka, J. Insect Physiol. 18(1972)1351-1367.
17. E. N. Lambremont, Comp. Biochem. Physiol. 41B(1972)337-342 .
18. A.M.T. Beenakkers, Insect Biochem. 3(1973)303-308.
19. J. J. Brown and G. M. Chippendale, J. Insect Physiol. 20(1974)1117-1130.

20. K. H. Tan, Comp. Biochem. Physiol. 46B(1973)1-7.

21. L. L. Bieber, E. Hodgson, V. H. Cheldelin, V. J. Brookes and R. W. Newburgh, J. Biol. Chem. 236(1961)2590-2595.

22. G. Y. Sun and V. J. Brookes, Comp. Biochem. Physiol. 24(1968) 177-185.

23. M. A. D'Costa and L. M. Birt, J. Insect Physiol. 12(1966)1377-1394.

24. P. R. Walker, L. Hill and E. Bailey, J. Insect Physiol. 16(1970) 1001-1015.

25. E. N. Lambremont, J. E. Bumgarner and A. F. Bennett, Comp. Biochem. Physiol. 19(1966)417-429.

26. J. K. Mauldin, E. N. Lambremont and J. B. Graves, Insect Biochem. 1(1971)316-326.

27. C. H. Schaefer, J. Insect Physiol. 14(1968)171-178.

28. M. Overturf and R. L. Dryer, in G. A. Kerkut (Editor), Experiments in the Biochemistry of Animal Lipids. In Experiments in Physiology and Biochemistry, Academic Press, New York, 1969, p. 113.

29. F. Chang, Comp. Biochem. Physiol. 49B(1974)567-578.

30. E. Stevenson, J. Insect Physiol 8(1972)1751-1756.

31. N. S. Radin, in J. M. Lowenstein (Editor), Methods in Enzymology, Vol. 14, in Preparation of Lipid Extract, Academic Press, New York, 1969, p. 245.

32. K. K. Thomas, J. Insect Physiol. 20(1974)845-858.

33. G. Rouser, G. Kritchevsky and D. Heller, J. Am. Oil Chem. Soc. 40(1963) 425-454.

34. E. G. Bligh and W. G. Dyer, Can J. Biochem. Physiol. 37(1959) 911-917.

35. J. S. Barlow, Can. J. Biochem. 42(1964)1365-1374.

36. M. P. Castillon, C. Jimenez, R. E. Catalan and A. M. Municio, Insect Biochem. 1(1971)309-315.

37. S. N. Thompson and R. B. Bennett, J. Insect Physiol. 17(1971)1555-1563.

38. S. N. Thompson and J. S. Barlow, Can. J. Zool. 50(1972)1033-1034.

39. A. B. Dutkowski, J. Insect Physiol. 19(1973)1721-1726.

40. R. F. N. Hutchins and M. M. Martin, Lipids 3(1968)247-249.

41. V. P. Dole and H. Meinertz, J. Biol. Chem. 235(1960)2595-2597.

42. F. Chang and S. Friedman, Insect Biochem. 1(1971)63-80.

43. M. A. Madariaga, A. M. Municio and A. Ribera, Comp. Biochem. Physiol. 36(1970)271-278.

44. M.A.B. Maxwell and J.P. Williams, J. Chrom. 31(1967)62-68.

45. G. W. K. Cavill and P. J. Williams, J. Insect Physiol. 13(1967)
    1097-1103.

46. G. W. K. Cavill, D. V. Clark, M.E.H. Howden and S. W. Wyllie,
    J. Insect Physiol. 16(1970)1721-1728.

47. G. R. Karuhize, Comp. Biochem. Physiol. 438(1972)563-569.

48. R. Kaschnitz, M. Peterlik and H. Weiss, Anal. Biochem. 30(1969)
    147-148.

49. L. I. Gilbert, Adv. Insect Physiol. 4(1967)69-211.

50. B. Borgström, Acta Physiol. Scand. 25(1952)101-110.

51. P. A. Marks, A. Gellhorn and C. Kidson, J. Biol. Chem. 235(1960)
    2579-2583.

52. J. J. Brown and G. M. Chippendale, J. Insect Physiol. 19(1973)607-
    614.

53. R. D. McCarthy and A. H. Duthie, J. Lipid Res. 3(1962)117-119.

54. T. L. Chang and C. C. Sweeley, Biochemistry 2(1963)592-604.

55. R. Wood and F. Snyder, Lipids 3(1968)129-135.

56. J. Hirsch and E. H. Ahrens, Jr., J. Biol. Chem. 233(1958)311-327.

57. K. K. Carroll, J. Lipid Res. 2(1961)135-141.

58. K. K. Carroll, J. Am. Oil Chem. Soc. 40(1963)413-419.

59. J. G. Durocher and B. Leboeuf, Can. J. Biochem. 47(1969)746-750.

60. G. Cavina, G. Moretti, A. Mollica, L. Moretta and P. Siniscalchi,
    J. Chrom. 44(1969)493-508.

61. N. T. Werthessen, J. R. Beall and A. T. James, J. Chrom. 46(1970)
    149-160.

62. O. S. Privett, M. L. Blank, D. W. Codding, and E. C. Nickell,
    J. Am. Oil Chem. Soc. 42(1965)381-383.

63. H. K. Mangold, J. Am. Oil Chem. Soc. 38(1961)708-718.

64. G. M. Chippendale, Insect Biochem. 1(1971)283-292.

65. Y. Peled and A. Tietz, Insect Biochem.5(1975)61-72.

66. L. J. Brown and J. M. Johnston, J. Lipid Res. 3(1962)480-481.

67. C. P. Freeman and D. West, J. Lipid Res. 7(1966)324-327.

68. J. L.Holjnaki and S. C. Smith, J. Chrom. 90(1974)365-367.

69. O. S. Privett and M. L. Blank, J. Am. Oil Chem. Soc. 39(1962)520.

70. M. E. Fewster, B. J. Burns and J. F. Mead, J. Chrom. 43(1969)
    120-126.

71. E. Van Handel and D. B. Zilbersmit, J. Lab.Clin. Med. 50(1957)
    152-157.

72. S. V. Pande, R. Parvin Khan and T. A. Venkitasubramanian, Anal.
    Biochem. 6(1963)415-423.

73. W. G. Duncombe, Biochem. J. 88(1963)7-10.

74. O. B. Lindsay and J. S. Barlow, Comp. Biochem. Physiol. 39B(1971)
    823-832.

75. L. D. Metcalfe and A. A. Schmitz, Anal. Chem. 33(1961)363-364.

76. L. D. Metcalfe, A. A. Schmitz and J. R. Pelka, Anal. Chem. 38
    (1966)514-515.

77. S. H. W. Cmelik, Insect Biochem. 2(1972)361-366.

78. S. A. Hyun, G. V. Vahouny and C. R. Treadwell, Anal. Biochem.
    10(1965)193-202.

79. W. R. Morrison and L. M. Smith, J. Lipid Res. 5(1964)600-608.

80. J. S. Martin, J. Insect Physiol. 15(1969)1025-1045.

81. D. Van Wijngaarden, Anal.Chem. 39(1967)848-849.

82. J. I. Peterson, H. deSchmertzing and K. Abel, J. Gas Chromatography
    3(1965)126-130.

83. W. K. Fulk and M. S. Shorb, J. Lipid Research 11(1970)276-277.

84. G. V. Marinette, Biochemistry 1(1962)350-353.

85. H. Schlenk and J. L. Gellerman, Anal.Chem. 32(1960)1412-1414.

86. L. L. Jackson, G. L. Baker and J. E. Henry, J. Insect Physiol. 14
    (1968)1773-1778.

87. J. G. Saha, R. L. Randell and P. W. Riegert, Life Sci. 5(1966)1597-
    1603.

88. M. Hoshi, M. Williams and Y. Kishimoto, J. Lipid Res. 14(1973)
    599-601.

89. T. E. Morgan, D. J. Hanahan and J. Ekholm, Fedn. Proc. 22(1963)
    414(Abstract).

90. P. Mareš, J. Chrom. 32(1968)745.

91. O. Renkonen, Acta Chem. Scand. 17(1963)634-640.

92. A. Luukkonen, M. B. Korvenkontio and O. Renkonen, Biochem.
    Biophys. Acta 326(1973)256-261.

93. E. N. Lambremont, Insect Biochem.2(1972)197-202.

94. M. E. Mason and G. R. Waller, Anal. Chem. 36(1964)583-586.

95. S. B. Tove, J. Nutrition 75(1961)361.

96. W. W. Christie, Analyst 97(1972)221-223.

97. V. L. Davison and H. J. Dutton, J. Lipid Res. 8(1967)147-149.

98. E. C. Horning, E. A. Moscatelli and C. C. Sweeley, Chem. Ind.
    (1959)751-752.

99. C. L. Sloan, L. L. Jackson, G. L. Baker and J. E. Henry, Lipids
     3(1968)455-456.

100. L. T. Kok and D. M. Norris, J. Insect Physiol. 18(1972)1137-1151.

101. S. N. Thompson and J. S. Barlow, Comp. Biochem. Physiol. 36(1970)
     103-106.

102. R. O. Henson, A. C. Thompson, R. C. Gueldner and P. A. Hedin,
     J. Insect Physiol. 18(1972)161-167.

103. A. T. James, in D. Glick (Editor), Methods in Enzymology, Vol. 8,
     Interscience, New York, N. Y., 1960, p.45.

104. B. M. Craig, in Gas Chromatography, Academic Press Inc. New York,
     N.Y., 1962.

105. M. T. Armold and F. E. Regnier, J. Insect Physiol. 21(1975)1827-
     1833.

106. K. K. Carroll, Nature 191(1961)377-378.

107. F. Snyder and B. Malone, Biochem. Biophys. Res. Commun. 41
     (1970)1382-1387.

108. C. B. Barrett, M. S. J. Dallas and F. B. Padley, J. Am. Oil Chem.
     Soc. 40(1963)580-584.

109. R. Wood and F. Synder, J. Am. Oil Chem. Soc. 43(1966)53-54.

110. J. A. Svoboda, M. J. Thompson and W. E. Robbins, Life Sci. 6
     (1967)395-404.

111. J. A. Svoboda, M. J. Thompson, T. C. Elden and W. E. Robbins,
     Lipids 9(1974)752-755.

112. J. Avigan, DeW. S. Goodman and D. Steinberg, J. Lipid Res. 4
     (1963)100-103.

113. J. N. Kaplanis, W. E. Robbins, H. E. Vroman and B. M. Bryce,
     Steroids 2(1963)547-553.

114. J. A. Svoboda, M. Womack, M. J. Thompson and W. E. Robbins,
     Comp. Biochem. Physiol. 30(1969)541-549.

115. K. C. Goodnight and H. W. Kircher, Lipids 6(1971)166-169.

116. H. E. Vroman, J. N. Kaplanis and W. E. Robbins, J. Lipid Res. 5
     (1964)418-421.

117. W. E. Robbins, R. C. Dutky, R. E. Monroe and J. N. Kaplanis,
     Ann. Entomol. Soc. Amer. 55(1962)102-104.

118. M. G. Horning, E. A. Williams and E. C. Horning, J. Lipid Res.
     1(1960)482-485.

119. M. M. Martin and G. A. Carls, Lipids 3(1968)256-259.

120. J. D. Johnston, F. Gautschi and K. Block, J. Biol. Chem. 224(1957)
     185-190.

121. P. R. Moore and C. A. Baumann, J. Biol. Chem. 195(1953)615-621.

122. W. M. Sperry and M. Webb, J. Biol. Chem. 187(1950)97-102.

123. W. Bergmann, J. Biol. Chem. 132(1940)471-472.

124. J. P. Allais, A. Alcaide, and M. Barbier, Experientia 29(1973)
944-945.

125. J. A. Svoboda, M. J. Thompson and W. E. Robbins, Lipids 7(1972)
156-158.

126. J. A. Svoboda, M. J. Thompson, W. E. Robbins and T. C. Elden,
Lipids 10(1975)524-527.

127. H. E. Nordby and S. Nagy, J. Chrom. 75(1973)187-193.

128. N. L. Lasser and R. B. Clayton, J. Lipid Res. 7(1966)413-421.

129. W. R. Roeske and R. B. Clayton, J. Lipid Res. 9(1968)276-284.

130. R. C. Dutky, W. E. Robbins, J. N. Kaplanis and T. S. Shortino,
Comp. Biochem. Physiol. 9(1963)251-255.

131. R. F. N. Hutchins and J. N. Kaplanis, Steroids 13(1969)605-613.

132. J. A. Svoboda, J. N. Kaplanis, W. E. Robbins and M. J. Thompson,
Ann. Rev. Entomol. 20(1975)205-220.

133. J. A. Svoboda, W. E. Robbins, C. F. Cohen and T. S. Shortino,
in J. G. Rodriguez (Editor), Insect and Mite Nutrition, North-Holland,
Amsterdam, 1972, p. 505.

134. R. D. Henson, A. C. Thompson, R. C. Gueldner and P. A. Hedin,
Lipids 6(1971)352-355.

135. H. D. Crone and R. G. Bridges, Biochem. J. 89(1963)11-21.

136. S. Turunen, J. Insect Physiol. 19(1973)2327-2340.

137. J. E. Kinsella, Comp. Biochem. Physiol. 17(1966)635-640.

138. C. M. Wang and R. L. Patton, J. Insect Physiol. 15(1969)851-860.

139. E. N. Lambremont, and R. Wood, Lipids 3(1968)503-510.

140. V. P. Skipski, R. F. Peterson and M. Barclay, Biochem. J. 90(1964)
374-378.

141. D. Abramson and M. Blecher, J. Lipid Res. 5(1964)628-631.

142. G. Rouser, C. Galli, E. Lieber, M. L. Blank and O. S. Privett,
J. Am. Oil Chem. Soc. 41(1964)836-840.

143. G. Rouser, G. Kritchevesky, C. Galli and D. Heller, J. Am. Oil
Chem. Soc. 42(1965)215-227.

144. G. Rouser, A. N. Siakotos and S. Fleischer, Lipids 1(1966)85-86.

145. G. Rouser, S. Fleischer, A. Yamamoto, Lipids 5(1970)494-496.

146. E. Y. Lipsitz and J. E. McFarlane, Comp. Biochem. Physiol. 34
(1970)699-705.

147. W. D. Skidmore and C. Entenman, J. Lipid Res. 3(1962)471-475.

148. C. M. VanGent, O. J. Roseleur and P. VanDerBijl, J. Chrom. 85 (1973)174-176.

149. N. M. Meskovic, J. Chrom. 27(1967)488-490.

150. L. J. Nutter and O. S. Privett, J. Chrom. 35(1968)519-525.

151. G. R. Bartlett, J. Biol. Chem. 234(1959)466-468.

152. G. V. Marinetti, J. Lipid Res. 3(1962)1-20.

153. J. S. Chahl, Experientia 27(1971)608-610.

154. A. F. Rosenthal and S. C. H. Han, J. Lipid Res. 10(1969)243-245.

155. K. D. P. Rao and H. C. Agarwal, Comp. Biochem. Physiol. 30(1969) 161-167.

156. A. R. Beaudoin and A. Lemonde, J. Insect Physiol. 16(1970)511-519.

157. H. M. Jenkin, E. McMeans, L. E. Anderson, Lipids 10(1975)686-694.

158. H. Chino and L. I. Gilbert, Biochim. Biophys. Acta 98(1965)94-110.

159. H. Chino, A. Sudo and K. Harashima, Biochim. Biophys. Acta 144 (1967)177-179.

160. H. Chino, S. Murakami and K. Harashima, Biochim. Biophys. Acta 176(1969)1-26.

161. K. K. Thomas and L. I. Gilbert, Arch. Biochem. Biophys. 127(1968) 512-521.

162. K. K. Thomas and L. I. Gilbert, Physiol. Chem. Phys. 1(1969)293-311.

163. E. Whitmore and L. I. Gilbert, Comp. Biochem. Physiol. 47B(1974) 63-78.

164. R. M. Zacharins, T. E. Zell, J. H. Morrison and J. J. Woodlock, Anal. Biochem. 30(1969)148-152.

165. P. Sroka and R. H. Barth, Insect Biochem. 5(1975)637-645.

166. M. Suzuki, M. Kobayashi and N. I. Kekawa, Lipids 5(1970)539-544.

167. J. A. Svoboda, J. H. Pepper and G. L. Baker, J. Insect Physiol. 12(1966)1549-1565.

168. J. E. Bumgarner and E. N. Lambremont, Comp. Biochem. Physiol. 18(1966)975-981.

169. E. J. Barron and D. J. Hanahan, J. Biol. Chem.231(1958)493-503.

170. J. E. McFarlane, Comp. Biochem. Physiol. 24(1968)377-384.

171. J. Lamond, Analyst 74(1949)560.

172. F. E. Regnier, Biol. Of Reprod. 4(1971)309-326.

173. W. P. Olson, Comp. Biochem. Physiol. 35(1970)273.

174. P. Gerolt, J. Insect Physiol. 15(1969)563.

175. L. L. Jackson and G. L. Baker, Lipids 5(1970)239-246.

176. L. L. Jackson and G. J. Blomquist, in P. E. Koluttukudy (Editor), Chemistry and Biochemistry of Natural Waxes, Elsevier, Amsterdam, 1976,Ch.6.

177. J. B. Lok, E. W. Cupp and G. J. Blomquist, Insect Biochem. 5 (1975)821-829.

178. G. J. Blomquist, C. L. Soliday, B. A. Byers, J. W. Brakke and L. L. Jackson, Lipids 7(1972)356-362.

179. C. L. Soliday, G. J. Blomquist and L. L. Jackson, J. Lipid Res. 15 (1974)399-405.

180. L. L. Jackson, Lipids 5(1970)38-41.

181. L. L. Jackson, Comp. Biochem. Physiol. 41B(1972)331-336.

182. L. L. Jackson, M. T. Armold and F. E. Regnier, Insect Biochem. 4 (1974)369-379.

183. D. L. Silhacek, D. A. Carlson, M. S. Mayer and J. D. James, J. Insect Physiol.18(1972)347-354.

184. B. S. Goodrich, J. Lipid Res. 11(1970)1-6.

185. D. R. Nelson and D. R. Sukkestad, J. Lipid Res. 16(1975)12-18.

186. A. R. Gilby and J. W. McKellar, J. Insect Physiol.16(1970)1517-1529.

187. M. T. Armold and F. E. Regnier, J. Insect Physiol. 21(1975)1581-1586.

188. M. T. Armold, G. J. Blomquist and L. L. Jackson, Comp. Biochem. Physiol. 31(1969)685-692.

189. R. B. Turner and F. Acree, Jr., J. Insect Physiol. 13(1967)519-522.

190. D. R. Nelson, D. R. Sukkestad and A. C. Terranova, Life Sci. 10 (1971)411-419.

191. P. A. Diehl, Nature 244(1973)468-470.

192. R. F. N. Hutchins and M. M. Martin, Lipids 3(1968)250-255.

193. M. M. Martin and J. G. MacConnell,Tetrahedron 26(1970)307-319.

194. G. L. Baker, H. E. Vroman and J. Padmore, Biochem. Biophys. Res. Commun. 13(1963)360-365.

195. I. M. Beatty and A. R. Gilby, Die Naturwissenschaften 56(1969)373.

196. W. S. Bowers and M. J. Thompson, J. Insect Physiol. 11(1965) 1003-1011.

197. K. S. Brown, Chem. Soc. Rev. 4(1975)263-288.

198. A. C. Chibnall, A. L. Latner, E. F. Williams and C. A. Ayre, Biochem. J. 28(1934) 313· 325.

199. A. C. Chibnall, S. H. Piper, A. Pollard, E. F. Williams and P. N. Sahai, Biochem.J. 28(1934) 2189.

200. E. Faurot-Bouchet and G. Michel, Bull. Soc. Chim. Biol. 47(1965) 93-97.

201. E. Faurot-Bouchet and G. Michel, J. Am. Oil Chem. Soc. 41(1964) 418-421.

202. L. Savidan, Annales Chimie 1(1956)53-83.

203. Y. Tamaki, Lipids 3(1968)186-187.

204. M. Kono and T. Maruyama, J. Agr. Chem. Soc. Japan 14(1938)318.

205. R. H. Hackman, Arch. Biochem. Biophys. 33(1951)150-154.

206. A. R. Gilby and A. E. Alexander, Arch. Biochem. Biophys. 67(1957) 302-306

207. A. R. Gilby, Arch. Biochem. Biophys. 67(1957)307-319.

208. T. Rios and F. Colunga, Chem. Ind. (1965)1184-1185.

209. Y. Iiteka, I. Watanabe, I. T. Harrison and S. Harrison, J. Am. Chem. Soc. 90(1968)1092-1093.

210. T. Rios and L. Quijano, Tetrahedron Letters 17(1969)1317-1318.

211. T. Rios and F. Gomez G., Tetrahedron Letters 34(1969)2929-2930.

212. T. Rios, L. Quijano, and J. Calderon, Chem. Comm. (1974)728-729.

213. T. Rios and S. Perez, Chem. Comm. (1969)214-215.

214. Y. Tamaki and S. Kawai, Appl. Ent. Zool. 4(1968)79-86.

215. Y. Tamaki, T. Yushima and S. Kawai, Appl. Ent. Zool.4(1969)126-134.

216. Y. Tamaki, Lipids 1(1966)297-300.

217. Y. Tamaki, Nogyo Gijutsu Kenkyusho Hokoku B(1970)1-111.

218. Y. Tamaki and S. Kawai, Jap. J. appl. Ent. Zool. 12(1968)23-28.

219. A. Hashimoto, A. Hirotani, K. Mukai and S. Kitaoka, Agr. Biol. Chem. 34(1970)1839.

220. A. Hashimoto and S. Kitaoka, Agr. Biol. Chem. 35(1971)275-277.

221. A. Hashimoto, H. Yoshida, K. Mukai and S. Kitaoka, J. Agr. Chem. Soc. Japan 45(1971)96-99.

222. A. Hashimoto, A. Hirotani, K. Mukai and S. Kitaoka, J. Agr. Chem. Soc. Japan 45(1971)100-109.

223. A. Hashimoto, A. Hirotani, M. Katsunori, K. Shozaburo, Nippon Nogei Kagaku Kaishi 45(1971)100-105.

224. H. Singh, R. Madhav, T. R. Seshadri and G. B. V. Subnamanian, Tetrahedron 23(1967)4795-4800.

225. M. S. Wadia, R. G. Khurana, V.V. Mhaskar and S. Dev, Tetrahedron 25(1969)3841-3854.

226. W. W. Christie, F. D. Gunstone and H. G. Prentice, J. Chem. Soc. (1963)5768.

227. R. G. Khurana, M. S. Wadia, V.V. Mhaskar and S. Dev, Tetrahedron Letters 24(1964)1537-1543.

228. R. Madhav, T. R. Seshadri and G. B. V. Subramanian, Indian J. Chem. 5(1967)182.

229. M. S. Wadia, V. V. Mhaskar and S. Dev, Tetrahedron Letters (1963)513-517.

230. A. N. Singh, A. B. Upadhye, M. S. Wadia, V. V. Mhaskar, and S. Dev, Tetrahedron 25(1969)3855-3867.

231. R. C. Cookson, N. Lewin and A. Morrison, Tetrahedron 18(1962) 547-558.

232. R. C. Cookson, A. Molera and A. Morrison, Tetrahedron 18(1962) 1321-1323.

233. R. G. Khurana, A. N. Singh, A. B. Upadhye, V. V. Mhaskar, and S. Dev, Tetrahedron 26(1970)4167-4175.

234. T. R. Seshadri, N. Sriram and G. B. V. Subramanian, Indian J. Chem. 9(1971)528.

235. A. B. Upadhye, M. S. Wadia, V. V. Mhaskar and S. Dev, Tetrahedron 26(1970)4177-4187.

236. A. B. Upadhye, M. S. Wadia, V. V. Mhaskar and S. Dev., Tetrahedron 26(1970)4387-4396.

237. J. W. White, M. L. Reithof and I. Kuchnir, J. Assoc. Off. Anal. Chem. 43(1960)781.

238. A. P. Tulloch and L. L. Hoffman, J. Am. Oil Chem. Soc. 49(1972) 696-689.

239. United States Pharmacopeia, Vol. XVIII, Mack Printing Co., Easton, Pa., 1970,p. 779.

240. J. W. White, Jr., M. K. Reader and M. L. Riethof, J. Assoc. Off. Anal. Chem. 43(1960)778-780.

241. P. J. Holloway, J. Am. Oil Chem. Soc. 46(1969)189-190.

242. P. Mazliak, Phytochemistry 2(1963)253-261.

243. A. P. Tulloch, Lipids 5(1970)247-258.

244. D. H. S. Horn, Z. H. Kranz and J. A. Lamberton, Australian J. Chem. 17(1964)464.

CHAPTER 6

CHEMICAL ANALYSIS OF INSECT MOLTING HORMONES

MASATO KOREEDA and BEVERLY A. TEICHER

Department of Chemistry, The Johns Hopkins University
Baltimore, Maryland 21218 (U. S. A.)

CONTENTS

INTRODUCTION

Recent remarkably rapid advances in the field of insect chemistry have allowed the elucidation of many fundamental biological phenomena on a molecular basis.   This is due mainly to
newly developed methods in organic spectroscopy and separation
techniques for dealing with micro-quantities which have opened
this field to the organic chemist.

The insect life cycle is known to be governed by the action
of several hormones (1).   Of these, the molting and juvenile
hormones, both of which are low molecular weight organic compounds, have been well studied chemically.   The molting hormone
of insects and crustaceans was the first among the many invertebrate hormones to be structurally elucidated.   Thus, the
structure of the molting hormone, α-ecdysone 1, which was first
isolated by Butenandt and Karlson from the silk-worm (Bombyx
mori) in crystalline form in 1954 (2), was proven by X-ray analysis without the use of a heavy atom in the molecule in 1965

α-ecdysone   β-ecdysone   ponasterone A

(3).   Subsequently other ecdysones were isolated from several
sources including crustaceans, i.e., ß-ecdysone $\underset{\sim}{2}$ (crustecdysone,
ecdysterone, 20-hydroxyecdysone) from crayfish (4) and silk-worm
(5).   Both of these two hormones and other structurally related
steroids have now been isolated from a number of invertebrates
(see Table I) (6).

The discovery of polyhydroxy steroids with molting hormone
activity in the plant kingdom (phytoecdysone) in 1966 (7,8) has
provided a good source of ecdysones, e.g., ponasterone A $\underset{\sim}{3}$ in
ca. 2% yield from Podocarpus nakaii leaves (9) vs. 0.007% yield
of α-ecdysone from silk-worm pupae.   In addition, the extraction procedures required for phytoecdysones are much simpler
than those required for zooecdysones (ecdysones from the animal
kingdom) (10).   There have been approximately forty phytoecdysones characterized thus far (6,11,12).   These have been used
for the study of the relationship of activity to structure, as
well as for more macroscopic biological and entomological studies.

This review deals with a brief description of the extraction
of ecdysones principally from insects or crustaceans, and with
the current status in the chemical analysis of ecdysones from
insects with forcus on some newer spectroscopic analysis of the
ecdysone structure.

EXTRACTION AND ISOLATION OF ECDYSONES

In general, due to the extremely small quantities of ecdysones in invertebrates ($10^{-9} \sim 10^{-6}$ g per animal), their isolation
inevitably requires laborious and lengthy procedures and starting with an enormous amount of the raw material.   The diffi-

Table I. List of the isolated zooecdysones and their sources.

ecdysones	sources
α-ecdysone 1	silkworm, tobacco hornworm, cockroach (Leucophaea)
β-ecdysone 2	silkworm, tobacco hornworm, oak-silk moth, nematode, crayfish, marine crab
2-deoxy-β-ecdysone 4	crayfish
26-hydroxy-α-ecdysone 5	tobacco hornworm
20,26-dihydroxy-α-ecdysone 6	tobacco hornworm
inokosterone (callinecdysone A) 7	marine crab
makisterone A(callinecdysone B) 8	marine crab, milkweed bug

Table II. Distribution coefficients (K values) by counter-current distribution and R_F values by thin-layer chromatography of some of the zooecdysones.

ecdysones		K^{*}	$R_F{}^{\#}$
α-ecdysone	1	3.54	0.23
makisterone A	8	1.27	0.20
β-ecdysone	2	0.52	0.15
3-epi-β-ecdysone	9	0.52	0.17
26-hydroxy-α-ecdysone	5	0.39	0.08
20,26-dihydroxy-α-ecdysone	6	0.06	0.05

* Countercurrent distribution: solvent system, cyclohexane, n-butanol, and water (5:5:10): 24-26°C.
\# Preparative silica gel G plates: solvent system, chloroform and ethanol (8:2) with wick.
Taken from reference (13).

culty in isolating ecdysones can be attributed to their considerable solubility in aqueous systems, α-ecdysone is an exception
to this behavior (see Table II) (13).   The original extraction
scheme from silk-worm pupae of Karlson et al. (14) seems to be
effective for the relatively less polar ecdysones (such as α-
ecdysone), however, in order to isolate the more polar ecdysones,
including β-ecdysone 2, this scheme had to be altered.   Chart
I summarizes the extraction scheme, without the use of alumina
chromatography, established by Horn et al. (15) for the isolation of β-ecdysone from crayfish.   This procedure, which minimizes the loss of ecdysones, can readily be applied to the isolation of polar ecdysones from insects.

Recently, the USDA group (13) reported a simpler and more
effective procedure for the isolation of ecdysones from the
developing embryo of the milkweed bug.   Their method involves
chromatographic purification of the concentrated crude methanol
extract on deactivated neutral alumina (Woelm), followed by
silica gel chromatography using a mixture of benzene and methanol as the eluting solvent.   Earlier, this same group (16)
developed a proficient method for the separation of ecdysone
mixtures using preparative silica gel G chromatographic plates
and a solvent system of varying ratios of chloroform and ethanol.   The R_F values of some typical zooecdysones are listed in
Table II.

An efficient separation of ecdysone mixtures was also effected
by paper chromatography utilizing a two-phase solvent system.
For example, a mixture of α- and β-ecdysone, ponasterone A, inokosterone, and makisterone A was separated using a solvent system consisting of benzene/iso-propanol/water, toluene/iso-propa-
nol

Frozen crayfish waste (1 ton!)

    Extracted with aq. EtOH (3860 l) and concentrated in vacuum.

Aq. extract (360 l, 35 kg dissolved solids)

    Extracted with hexane:i-PrOH (200 l, 1:3).
    → Inactive pet. ether extract (110 l, 7 kg lipids).
    ← i-PrOH:water (1:1) backwash (55 l).

Defatted aq. extract (515 l)

    Ammonium sulfate (100 kg) added and extracted twice with hexane:i-PrOH
    (1:3, total of 360 l). combined hexane:i-PrOH layers concentrated in
    vacuum.

Aq. extract (4.5 l, 1.3 kg solids)

    Countercurrent extraction using n-BuOH:water, upper layer retained
    and concentrated.

Active extract (400 g activity: 10 Calliphora units (C.U.)/mg).

    Countercurrent extraction using $CHCl_3$:MeOH:water (1:2:1).
        Inactive $CHCl_3$ extract.
    MeOH layers retained and evaporated.

Active extract (71 g)

    Countercurrent extraction with $CHCl_3$:EtOH:aq. $KHCO_3$.
    $CHCl_3$ layer retained and evaporated.

Active extract (12 g)

    Reverse phase partition chromatography using n-BuOH:water.

    → 2-Deoxy-β-ecdysone fractions.

β-Ecdysone fractions (900 mg, activity: 100-200 C.U./mg)

    CM-Sephadex chromatography.

β-Ecdysone fractions (14.6 mg)

    Silicic acid chromatography.

β-Ecdysone (2.3 mg, activity: 40,000 C.U./mg)

Chart I.   Extraction scheme of β-ecdysone from crayfish by
          Horn et al. (15). Reproduced from reference (6)
          by permission of the copyright holders.

nol/water, or toluene/2-butanol/water.   The benzene/iso-propa-
nol/water (45% water) system, in particular, has excellent load
capacity and can be effectively used to purify quite crude material on Whatman No.3 MM paper (17).

Extraction of ecdysones from plants is generally much easier
because of the relatively high concentration of ecdysones in
plants.   However, one often faces a complex mixture of closely
related ecdysones.   This diiffcult problem has now been solved
by the use of high pressure liquid chromatography.

HIGH-PRESSURE LIQUID CHROMATOGRAPHY OF ECDYSONES

Although the separation and isolation of ecdysones can be
carried out effectively by normal phase silica gel chromatography, column or thin-layer, neither of these methods is suitable for quantitative analysis of ecdysones.   Ecdysones are
very polar molecules, therefore the solvent employed in normal
phase silica gel separation methods must be very polar.   In
thin-layer plate separations these solvents often elute some
binding material from the plate together with the steroids,
even when the plates have been pre-washed with the eluting solvent.   Furthermore, recovery of ecdysones from preparative
thin-layer plates and silica gel columns does not appear to be
quantitative in most cases.   This problem is particularly serious during the purification of polar metabolites of ecdysones.

The use of reverse phase high-pressure liquid chromatography
(HPLC) has eliminated the drawbacks inherent to both thin-layer
and column chromatography.   Hori (18) introduced HPLC to the
ecdysone field very early.   He showed that a reverse phase
HPLC system utilizing Amberlite XAD-2 resin and a water-ethanol

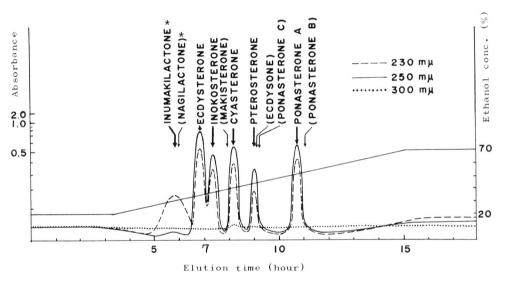

Fig. 1. HPLC of an authentic sample mixture of ecdysones with Amberlite XAD-2. A glass column(9 x 1600 mm) with a flow rate of 1 ml/min was used. * denotes a norditerpene which is often found in the extract of <u>Podocarpus</u> species. Reproduced from M. Hori, Steroids, 14(1969) 33, by permission of the copyright holders.

solvent gradient can be very effective in the purification of
ecdysones.   Amberlite XAD-2, an adsorption resin of the macro-
reticular type styrene-divinylbenzene copolymer, is available
from Rohm & Haas Co. (Philadelphia, Pa.); however, it is not
sold in the 200-400 mesh size used by Hori and must be pulverized to a finer uniform size.

A fair amount of very polar impurities, such as sugars, gly-
cosides and carboxylates, accompany ecdysones throughout the
conventional purification process.   These very polar impurities are removed quickly from reverse phase HPLC columns, then
the ecdysones are eluted.   A variety of ecdysones differing
only in the number of hydroxyl groups or their locations can
be well resolved by this method as shown in Fig. 1.   The strong
extinction coefficient of the 7-en-6-one chromophore in ecdysones makes HPLC capable of detecting a few gamma of ecdysones
by following the UV absorption at 254 nm.   The Amberlite XAD-
2 column with dimensions 36 x 1500 mm has a loading capacity
such that thirty to fifty milligrams of an ecdysone mixture
can be separated.

The most remarkable demonstration of the versatility of this
method is that injection of a crude methanol extract of Achyran-
thes fauriei showed the presence of both β-ecdysone and inokosterone as 0.32% and 0.25% of the root contents, respectively
(without any purification!) (Fig. 2).   Errors of the quantitative analysis which were obtained from integration of the peaks
were shown to be less than 2% using authentic ecdysone samples.
In contrast to plants, the concentration of ecdysones in invertebrates is quite low and usually it is not possible to identify ecdysone peaks from a crude extract of invertebrate even

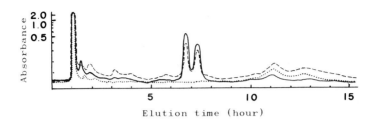

Fig. 2. HPLC of the crude extract of <u>Achyranthes fauriei</u> with Amberlite XAD-2. Reproduced from M. Hori, Steroids, 14(1969) 33, by permission of the copyright holders.

using reverse phase HPLC because of overlap with other UV absorbing materials.   It is necessary to fractionate invertebrate extracts either by silica gel column chromatography or by solvent-solvent extractions prior to purification by HPLC.

The isolation and analysis of ecdysones from invertebrates has been tremendously improved and simplified by the advent of HPLC.   Moriyama et al. (19) employed this method of detecting ecdysones very effectively in their study of the metabolites of α-ecdysone in the silk-worm.   As shown in Fig. 3, the formation of polar metabolites of 23,23,24,24-tetratritiated α-ecdysone, that is A, B, C, and β-ecdysone, was clearly demonstrated in the body fluid of 6-day old silk-worms.   The most polar metabolite A, is assumed to be a salt of isocaproic acid. Since metabolite C was hydrolyzed back to β-ecdysone by base, it was suggested that C is a polar ester of β-ecdysone (for example, a sulfate ester).

Although the HPLC method with Amberlite XAD-2 as developed by Hori is extremely useful, it suffers from one practical inconvenience, that is, the time required for one analysis is

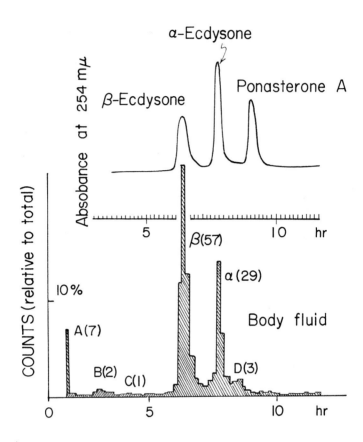

Fig. 3. Radiochromatogram of ecdysones and metabolites in day
6 Bombyx mori body fluid. The upper tracing shows the elution
pattern of standard ecdysones added to the extract (detected
by monitoring UV absorption at 254 nm). The lower bar graph
shows counts (relative to total) of fractions collected during
10.min intervals plotted against time. Compounds A, B, C, D,
and α- and β-ecdysones are noted; numbers in parentheses denote
percentage of total count of each compound. Redrawn from H.
Moriyama, et al., Gen. Comp. Endoclinol., 15(1970) 80, by permission of the copyright holders.

over 10 hrs.   Introduction of the use of Poragel PN (Waters

Associates, Inc., Milford, Mass.) as a column support has

solved this problem.   Poragel PN (supplied in a 37-75 μ size)

is a porous styrene-type polymer gel modified to introduce some

hydrophilicity and is designed for reverse phase separations.

Poragel PN retains the convenient features of Amberlite XAD-2,

but less than 60 minutes are required for most ecdysone ana-

lyses.   Recent refinements of both detector and column supports

have made possible the detection of as little as 0.1μg of ecdy-

sone, provided that impurities of similar polarity have been

removed.   It is advisable to use a water-methanol solvent

gradient, in order to maximize the number of theoretical plates,

when dealing with relatively non-polar ecdysones such as pona-

sterone A.   As with Amberlite XAD-2, Poragel PN can convenient-

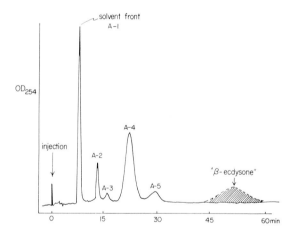

Fig. 4. HPLC of the crude fat body factor of the wax moth with
Waters ALC-100 (Poragel PN support, 3 ft x 3/8 in. od column,
50% methanol in water, with a 2.1 ml/min flow rate).   The
shaded β-ecdysone peak is from a separate 10 μg injection.
Reproduced from J. Benson, et al., Wilhelm Roux' Archiv, 175
(1974) 327, by permission of the copyright holders.

ly be used for the analysis and isolation of polar components.
As illustrated in Fig. 4, an active crude fraction of a fat
body factor,which stimulates evagination of wax moth Galleria
mellonella wing disks, in vitro, obtained from preparative silica
gel TLC, was further purified by Poragel PN chromatography (20).
In spite of speculation that this fat body factor may be either
an ecdysone or ecdysone analog, none of the major peaks, A-1∼
A-5 showed the characteristic spectral properties of ecdysones.
The structure of this fat body factor remains to be determined.
The review by Schooley and Nakanishi (21) presents a more detailed discussion on the use of Poragel PN.

Normal phase HPLC has been shown to be quite versatile in
the separation of relatively less polar ecdysones. Utilizing
Corasil II (37-50 μ, Waters Associates, Inc.), the USDA group
(22) showed that the ecdysone-like compounds with two to six
hydroxyl groups (ecdysterols) can be separated quickly. Corasil II, a column support made from core glass with two layers
of silica gel coating, can give a very high number of theoretical plates. A well resolved separation of typical ecdysterols
is reproduced in Fig. 5. This HPLC technique can be well
adapted to the analysis of biosynthetic intermediates of ecdysones from cholesterol or 7-dehydrocholesterol. The USDA group
(22) has proposed a general procedure for analyzing or isolating
specific ecdysterols from the crude ecdysterol mixtures obtained
from in vivo or in vitro systems. A partial purification by
column chromatography (possibly on silica gel) to obtain apolar,
polar, and more polar fractions is carried out prior to HPLC
separation. The solvent system employed in the HPLC depends
upon the polarity of the fraction to be analyzed.

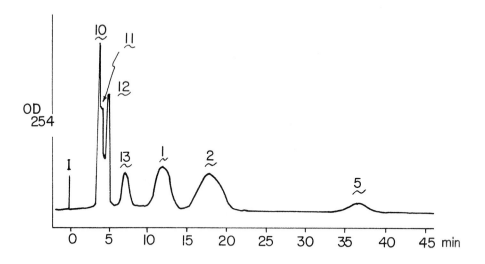

Fig. 5. HPLC of ecdysones and related steroids.
Column support: Corasil II (37-50 µ); column: three stainless steel U-shaped columns, 1.0 m x 2.0 mm i.d.; Instrument: DuPont 830 LC (uv detector at 254 nm); solvent:
chloroform/ethanol 9/1; flow rate: 1.5 ml/min; column
pressure: 1500 psi. I = injection. 1; α-ecdysone, 2; β-
ecdysone, 5; 26-hydroxy-α-ecdysone. Redrawn from H.N.
Nigg, et al., Steroids, 23(1974) 507, by permission of
the copyright holders.

Although the USDA group did not indicate the smallest amount
of ecdysones which it is possible to detect by this method,
taking into account the recent remarkable developments in the
technology of column packings and UV detectors, it is conceiv-
able that these ecdysterols can be analyzed in amounts as low
as ten nanograms.

Before closing the section on LC, it must be emphasized that
one of the advantages of separating ecdysones by LC is that no
derivatization is necessary.   The ecdysones are recovered from
LC column fractions by evaporating organic solvents under reduced pressure and lyophilization of aqueous solvents (after removal of most of organic solvents), thus minimizing any thermal
decomposition of unstable ecdysones.

GAS CHROMATOGRAPHIC ANALYSIS OF ECDYSONES

The extensive investigations by Horning's group (23) on the
use of gas chromatography (GC) as a means of analyzing polar
steroids such as corticosteroids and their metabolites proved
that GC can be effectively employed for analysis of these polar
steroids as their trimethylsilyl (TMS) ether or methoxime-tri-
methylsilyl ether derivatives.   They also showed that the
degree of silylation on the hydroxyls in polyhydroxy steroids
depends upon the silylating reagent used.   Since the amount of
sample required for GC analysis is extremely minute, in spite
of possible complications in the derivatization reaction, GC
remaines the most attractive means of detecting micro or sub-
micro quantities of ecdysones.   The first report on the use
of GC for ecdysone analysis by Katz and Lensky (24) demonstrated
its great potential.   β-Ecdysone, silylated by bis(trimethyl-

silyl)acetamide (BSA) (the structure of the derivative was not

given), was analyzed by GC with 3% SE-30 on Gas-Chrom Q with a

flame ionization detector (FID) (column temperature 250°C).

This method can detect 50 ng of α-ecdysone derivative.   Morgan

and Woodbridge (25) tried to find an ecdysone derivative which

would be volatile enough for GC and formed in high yield.   They

treated either α- or β-ecdysone with O-methylhydroxylamine

hydrochloride in pyridine at room temperature for 100 hrs; this

afforded almost quantitatively the two isomeric O-methyloximes.

The O-methyloxime derivative prevents formation of an enol sil-

ylate of the 7-en-6-one system and possible complication in the

enol silylation, since this reaction does not go to completion

readily.   The O-methyloximes were separated by preparative TLC

and each isomer was treated with BSA yielding silylated deriva-

tives of α- and β-ecdysone, 14 and 15, respectively.   It was

shown by mass spectrometry that
the teriary hydroxyls at C-14 and
20 were not silylated.   The four
silylated derivatives (a syn and
anti O-methyloxime mixture) of α-
and β-ecdysones were analyzed by
GC using five foot silylated glass

14 R=H

15 R=OH

columns (1% QF-1 or 3% OV-17 on CQ) at 232°C, with a nitrogen

flow rate of 50 ml/min. This same method was applied to partially

purified extracts of 5th instar nymphs of the desert locust

Schiostocerca gregaria.   Examination of groups of 10 nymphs

of S. gregaria taken at different ages in the 5th instar re-

vealed the presence of almost no α-ecdysone, but β-ecdysone was

detected with titers varying from 12 ng to 240 ng per insect.

The most systematic study published on the versatility of
GC in the ecdysone field is the work by Ikekawa's group (26).
There are two major problems associated with the derivatization
of ecdysones for GC, the presence of the 7-en-6-one system
which can be derivatized to an enol silylate under forcing conditions and the presence of sterically hindered hydroxyl groups.
By careful investigation, Ikekawa's group found conditions in
which the silylation reaction provides a single clean product
for each ecdysone.   With trimethylsilylimidazole (TSIM) at

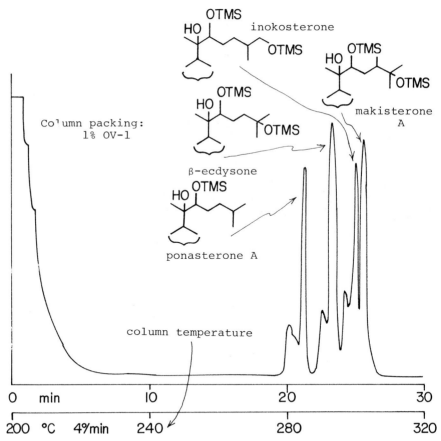

Fig. 6. GC separation of partially silylated TMS derivatives
of ecdysones.   By courtesy of Professor N. Ikekawa.

istic of β-ecdysone-TMS), which were formed by fission at C-20/

C-22 followed by hydrogen migration and elimination of TMS-OH,

respectively.   Fully silylated derivatives were chosen since

they are much more stable than partially silylated derivatives

or TMS-HFB derivatives.   This method constitutes the most pre-

cise and specific way of determining ecdysones.   Cyasterone

TMS derivative with an m/e 561 ion provides a standard.   The

calibration curves for the m/e 564 for α-ecdysone and m/e 561

for β-ecdysone are shown in Fig. 9.   This method is not only

capable of detecting ecdysones at the picogram level, it has

also allowed detection of a few previously unknown ecdysones

in insects (see Fig. 10).   Probably the most successful appli-

cation of mass fragmentography to ecdysone chemistry is the

study of the biosynthesis of α-ecdysone by the prothoracic

glands (28).   The long standing problem of the location of

ecdysone biosynthesis was resolved using this method.   Incu-

bation of isolated prothoracic glands of the Bombyx silk-worm

produced only α-ecdysone which was analyzed by HPLC (Zorbax

SIL) and most definitely by mass fragmentography, after being

purified by preparative silica gel TLC (Fig. 11).

In connection with these GC studies, Ikekawa's group (27)

developed a simple, efficient procedure for the isolation of

ecdysones from insect extracts.   Extraction from the crude

mixture with THF (Soxhlet) at its boiling point provided the

most satisfactory recovery of β-ecdysone; this was verified by

addition of a known amount of ^{3}H-β-ecdysone to the silk-worm

pupae, the recovery after extraction with THF of the ^{3}H-β-ecdy-

sone was essentially quantitative (98.5%).   The THF extract

was adsorbed on Carplex No. 80 (silica gel) and extracted suc-

cessively with n-hexane, benzene, and ether (Soxhlet).   This
removed 95% of the lipids, while the ecdysone remained in the
residue.   The ecdysones were extracted from the adsorbent with
THF, and the THF solution was passed through a short column of
silica gel.   The THF eluate was pure enough for analysis by GC
or by mass fragmentography using a GC-MS system.   This extraction procedure is superior to any previous sequences because of
its nearly complete recovery of ecdysones and its simplicity.

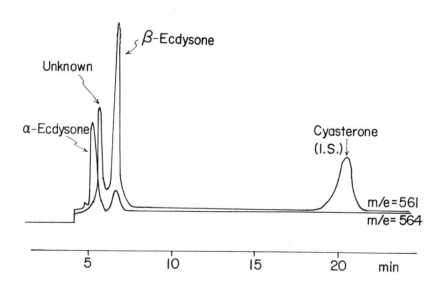

Fig. 10. Mass fragmentogram of ecdysones from 2
day-old silk-worm pupae. Redrawn from H. Miyazaki,
et al., Anal. Chem., 45(1973) 1164, by permission
of the copy right holders.

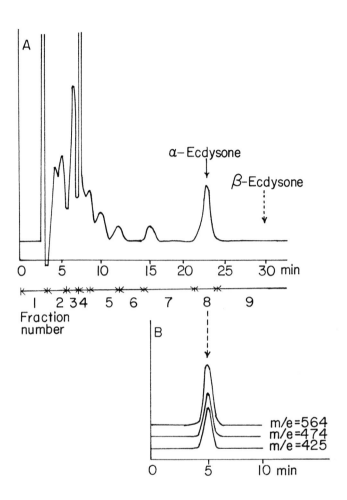

Fig. 11. (A) HPLC of the crude ecdysones biosynthesized by prothoracic glands after partial purification by preparative TLC. HPLC: instrument, DuPont 830 LC; column, Zorbax SIL (25 cm by 2.1 mm id); flow rate, 0.3 ml/min; column pressure, 1000 psi; column temperature, 20°C; and detector, ultraviolet photometer at 254 nm. (B) Mass fragmentogram of TMS derivative of fraction 8. Instrument, LKB 9000s MID PM 9060S; column, 1.5% OV-101 on Chromosorb W HP 80 to 100 mesh (1 mm by 4 mm id); column temperature, 270°C; ionization current, 60 μa; ionization voltage, 20 eV; accelerating voltage, 3.5 kv; and ion source temperature, 290°C. Redrawn from H. Chino, et al., Science, 183(1974) 529, by permission of the copyright holders.

RADIOIMMUNO-ASSAY OF ECDYSONES

Radioimmuno-assay (RIA) is a powerful technique for determining small amounts of steroid hormones (29).   This technique
takes advantage of the specificity and high binding affinity
of antibodies to detect the molecule in question.   The steroid
hormone is covalently bound to a carrier protein, usually bovine
serum albumin (BSA), thus it becomes haptenic and will elicit
a specific response from an immune system.

An antibody prepared by such a synthetic hapten-protein conjugate will form a complex with the free hapten.   However, because there is no possibility of cross-linking, the very typical
immune precipitate seen in antisera reactions with protein anti-
gens does not occur.   Nevertheless, it is possible to separate
the antibody-hapten complex from free hapten by a number of
methods.   If a radiolabel steroid is used, the percentage of
steroid bound by the antibody or remaining free can be determined by standard scintillation techniques.

An assay system based upon this principle typically uses an
amount of antiserum which will bind about 50% of the radiolabeled hapten.   Assuming that the radiolabeled hormone reacts
with the antibody similarly to the unlabeled hormone, it is
clear that increasing amounts of unlabeled hormone will dilute
the specific activity of the total hapten pool, and that progressively fewer counts will appear in the hapten-antibody complex (or conversely, progressively more counts will appear in
the free hapten pool).   The sensitivity of such a binding assay
is limited by the specific activity of the radiolabeled compound
and the binding affinity of the antibody.   The commercial
availability of some steroids with a specific activity greater

than 50 Ci/mmole and the production of antisera with binding

affinities of 10^9 and higher, have allowed measurements of less

than 0.05 picomoles of some steroids.

The use of RIA for the detection of β-ecdysone was first re-

ported by Borst and O'Connor (30).   β-Ecdysone was treated

with aminooxyacetic acid to give the oxime acetic acid (pre-

sumably a syn-anti mixture), which was converted to the mixed

anhydride and treated with BSA.   The protein-ecdysone conjugate

was dialyzed and lyophilized.   Three New Zealand white rabbits

were injected with this conjugate.

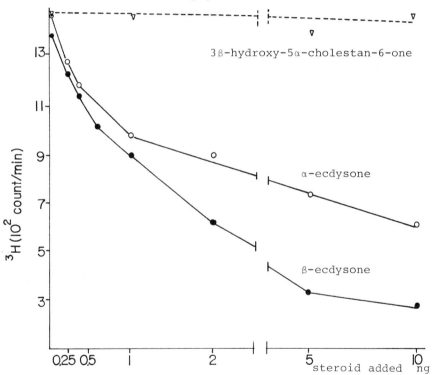

Fig. 12. Inhibition of ^{3}H-β-ecdysone binding by rabbit antiserum in the presence of increasing amounts of various unlabeled steroids. All reaction tubes contained 1% antiserum and ^{3}H-β-ecdysone (4000 cpm). Ordinate: cpm of labeled hapten bound by antiserum.   Redrawn from D.W. Borst
and J.D. O'Connor, Science, 178(1972) 418, by permission
of the copyright holders.

The specificity of the antibodies obtained by this method
was examined by RIA competition experiments; that is, the inhibition of radioactive ^{3}H-β-ecdysone binding by rabbit antiserum
was measured in the presence of increasing amounts of various
unlabeld steroids (Fig. 12).

They found that α-ecdysone and inokosterone would compete for
binding with ^{3}H-β-ecdysone, but were not as effective as β-ecdysone itself (Table V).   This means that this RIA method can be
used to detect α-ecdysone as well as β-ecdysone, though identification as α-ecdysone must be made by other means.

This method can reliably detect 200 picogram of β-ecdysone
or it has 25 times the maximum sensitivity of the bioassay.

RIA was effectively used to detect α-ecdysone in the media
secreted from a culture of prothoracic glands of tobacco hornworm, Manduca sexta (32).   Recently, Hagedorn et al. (33) reported that the ovary is a source of α-ecdysone in an adult mosquito, Aedes agypti (cultured in vitro); RIA played a major role

Table V. Relative Binding Efficiency of Ecdysone Analogs (31).

Steroids	labeled ecdysone used	
	β-ecdysone (6Ci/mmol)	α-ecdysone (50Ci/mmol)
β-ecdysone 2	100	100
α-ecdysone 1	75	71
2-deoxy-β-ecdysone 4	–	100
inokosterone 7	63	–
ponasterone A 3	1.1	–
22,25-dideoxy-α-ecdysone	1	–
cholesterol	0.004	–

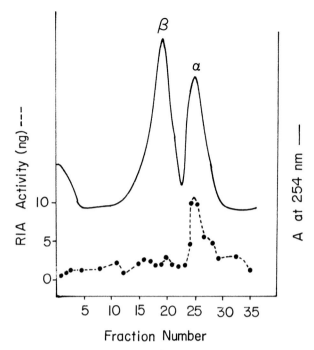

Fig. 13. Isocratic elution of RIA activity, and α- and β-ecdysone standards, from a reverse phase HPLC column. ————, absorbance at 254 nm of ecdysone standards;–●––●–, RIA activity of extracted medium partially purified using TLC. Redrawn from H.H. Hagedorn, et al., Proc. Nat. Acad. Sci., USA, 72(1975) 3255, by permission of the copyright holders.

in identifying α-ecdysone (see Fig. 13).   Two other groups (34,35) reported similar RIA methods for β-ecdysone.

CARBON-13 NUCLEAR MAGNETIC RESONANCE OF ECDYSONES

Among the many recently developed spectroscopic methods, Fourier transform carbon-13 nuclear magnetic resonance (FT-NMR) spectroscopy has a great potential for following the biosynthetic and metabolic fate of steroids (36).   By use of a microprobe for C-13, it has become possible to obtain a C-13 nmr spectrum of a sample of 1 mg or less, with C-13 present in natural abundance (1%).   If the compound were enriched in C-13

(2% or more) on a specific carbon, the sample size required for a

spectrum will accordingly decrease.   The C-13 nmr spectrum of

β-ecdysone (15 mg, with normal probe) in C_5D_5N is shown in Fig.

14.   The assignment of each peak is based on the report by

Hikino et al. (37).   All 27 carbons appear separately except

for C-2 and C-3.   This method has obvious advantage over the

proton nmr spectrum where most of the protons are unresolved

between δ 1.5 - 2.2 ppm.   The advantages inherent in C-13 nmr

can be utilized effectively in biosynthetic and metabolic stud-

ies of ecdysones at the semi-micro level if a specific carbon

is enriched in C-13.   Needless to mention, the structures of

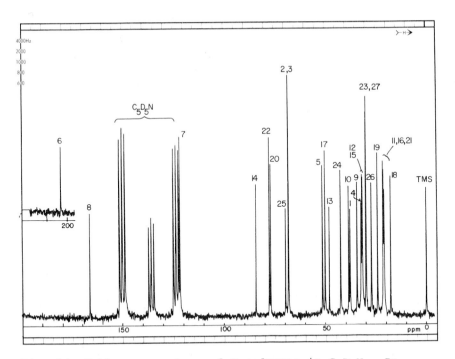

Fig. 14. C-13 nmr spectrum of β-ecdysone in C_5D_5N.   Recorded at the Johns Hopkins University   by M. Koreeda
using a Varian CFT-20 nmr spectrometer.

new ecdysones isolated from natural sources in amounts of 1 mg
can easily be analyzed by C-13 nmr (with a micro-probe).

SUMMARY

For a long time, the most fundamental and well established
means of assaying for ecdysone activity has been the bioassay.
A positive result can be obtained in the bioassay with as little
as 5 ng of ecdysones, depending upon the size of the test animal.   However, since individuals vary in their responsiveness,
it is necessary to use many specimens to determine an average
response.   Normally 10-30 animals are used per concentration,
thereby decreasing the actual sensitivity of the test to 50-
150 ng.   One advantage of the bioassay is that highly purified
samples are not required for the analysis unlike chemical analyses.   However, the problems of rearing a large number of
insects under proper conditions and the relative insensitivity
of this method are obvious drawbacks of the bioassay.

In contrast, GC analysis, especially with the electron capture detector system and mass fragmentography, and radioimmuno-
assay are both much more sensitive (to picogram levels) and are
more reliable in identifying specific ecdysones.   For the isolation of ecdysones, high-pressure liquid chromatography is by
far the most convenient technique when combined with partial
purification prior to liquid chromatography.

ACKNOWLEDGMENTS

The authors are indebted to Professor Koji Nakanishi (Colum-
bia University) who suggested that they write this review and
for his encouragement in this endeavor.   Thanks are due to

Mrs. Koko Koreeda who typed this manuscript and drew the figures and structures.

Finally, MK would like to acknowledge the financial support of the NIH (AI 12150).

REFERENCES

(1) a. L.I. Gilbert, in M. Rockstein (Editor), Physiology of
     Insects, Academic Press, New York, 1964, p.149; b. G.R.
     Wyatt, in G. Litwack (Editor), Biochemical Actions of Hor-
     mones, Vol. II, Academic Press, New York, 1972, Ch. 12, p.
     385; c. W.W. Doane, in S.J. Counce and C.H. Waddington (
     Editors), Developmental Systems: Insects, Vol. 2, Academic
     Press, New York, 1973, Ch. 4, p. 291: d. L.I. Gilbert and
     D.S. King, in M. Rockstein (Editor), The Physiology of In-
     secta, Academic Press, New York, 2nd ed., 1973, Ch.5, p.249.

(2) A. Butenandt and P. Karlson, Z. Naturforsch., 9b(1954) 389.

(3) W. Hoppe and R. Huber, Chem. Ber., 98(1965) 2353.

(4) F. Hampshire and D.H.S. Horn, Chem. Commun., (1966) 37.

(5) a. P. Hocks and R. Wiechert, Tetrahedron Lett., (1966)
     2989; b. H. Hoffmeister and H.F. Grützmacher, ibid.,
     (1966) 4017.

(6) D.H.S. Horn in M. Jacobson and D.G. Crosby (Editors), in
     Naturally Occurring Insecticides, Marcel Dekker, New York,
     1971, Ch. 9, p. 333.

(7) K. Nakanishi, M. Koreeda, S. Sasaki, M.L. Chang, and H.Y.
     Hsu, Chem. Commun., (1966) 917.

(8) M.N. Galbraith and D.H.S. Horn, Chem. Commun., (1966) 905.

(9) K. Nakanishi, M. Koreeda, M.L. Chang, and H.Y. Hsu, Tetra-
     hedron Lett., (1968) 1105.

(10) S. Imai, S. Fujioka, K. Nakanishi, M. Koreeda, and T. Kuro-
     kawa, Steroids, 10(1967) 557.

(11) H. Hikino and Y. Hikino, in W. Herz, H. Grisebach, and A.I.
     Scott (Editors), Progress in the Chemistry of Organic Nat-
     ural Products, Vol. XXVIII, Springer-Verlag, Wien and New
     York, 1970, p. 256.

(12) K. Nakanishi, M. Koreeda, and S. Imai, in S. Rangaswami
     and N.V. Subba Rao (Editors), Some Recent Developments
     in the Chemistry of Natural Products, Prentice Hall of

India, New Delhi, 1972, Ch. 13, p. 194.

(13) J.N. Kaplanis, S.R. Dutky, W.E. Robbins, M.J. Thompson,
     E.L. Lindquist, D.H.S. Horn, and M.N. Galbraith, Science,
     190(1975) 681.

(14) P. Karlson, H. Hoffmeister, W. Hoppe, and R. Huber, Justus
     Lieb. Ann. Chem., 662(1963) 1.

(15) D.H.S. Horn, S. Fabbri, F. Hampshire, and M.E. Lowe,
     Biochem. J., 109(1968) 399.

(16) J.N. Kaplanis, W.E. Robbins, M.J. Thompson, and S.R.
     Dutky, Science, 180(1973) 307.

(17) M.W. Gilgan and T.E. Farquharson, Steroids, 22(1973) 365.

(18) M. Hori, Steroids, 14(1969) 33.

(19) H. Moriyama, K. Nakanishi, D.S. King, T. Ohtaki, J.B.
     Siddall, and W. Hafferl, Gen. Comp. Endocrinol., 15(1970)
     80. See also: K. Nakanishi, H. Moriyama, T. Okauchi, S.
     Fujioka, and M. Koreeda, Science, 176(1972) 51.

(20) J. Benson, H. Oberlander, M. Koreeda, and K. Nakanishi,
     Wilhelm Roux' Archiv, 175(1974) 327.

(21) D.A. Schooley and K. Nakanishi, in E. Heftmann (Editor),
     Modern Methods of Steroid Analysis, Academic Press, New
     York, 1973, Ch. 2, p. 37.

(22) H.N. Nigg, M.J. Thompson, J.N. Kaplanis, J.A. Svoboda,
     and W.E. Robbins, Steroids, 23(1974) 507.

(23) a. E.C. Horning, M.G. Horning, N. Ikekawa, E.M. Chambaz,
     P.I. Jaakommaki, and C.J.M. Brooks, J. Gas Chromatog., 5
     (1967) 283; b. E.M. Chambaz and E.C. Horning, Anal. Lett.,
     1(1967) 201; c. M.G. Horning, A.M. Moss, and E.C. Horning,
     Analyt. Biochem., 22(1968) 284; d. E.M. Chambaz and E.C.
     Horning, ibid., 30(1969) 7; e. N. Sakauchi and E.C. Horn-
     ing, Anal. Lett., 4(1971) 41.

(24) M. Katz and Y. Lensky, Experientia, 26(1970) 1043.

(25) E.D. Morgan and A.P. Woodbridge, Chem. Commun., (1971)475.

(26) N. Ikekawa, F. Hattori, J. Rubio-Lightbourn, H. Miyazaki,
     M. Ishibashi, and C. Mori, J. Gas Chromatogr. Sci., 10
     (1972) 233.

(27) H. Miyazaki, M. Ishibashi, C. Mori, and N. Ikekawa, Anal.
     Chem., 45(1973) 1164.

(28) H. Chino, S. Sakurai, T. Ohtaki, N. Ikekawa, H. Miyazaki,
     M. Ishibashi, and H. Abuki, Science, 183(1974) 529.

(29) a. G.E. Abraham, in E. Heftmann (editor), Modern Methods

of Steroid Analysis, Academic Press, New York, 1973, Ch.
21, p. 451; b. C.J. Migeon, A. Kowarski, I.Z. Beitins,
and F. Bayard, ibid., Ch. 22. p. 471.

(30) D.W. Borst and J.D. O'Connor, Science, 178(1972) 418.

(31) D.W. Borst, Ph. D. Thesis, University of California,
Los Angeles, 1973, p. 50.

(32) D.S. King, W.E. Bollenbacher, D.W. Borst, W.V. Vedeckis,
J.D. O'Connor, P.I. Ittycheriah, and L.I. Gilbert, Proc.
Nat. Acad. Sci., USA, 71(1974), 793.

(33) H.H. Hagedorn, J.D. O'Connor, M.S. Fuchs, B. Sage, D.A.
Schlaeger, and M.K. Bohm, Proc. Nat. Acad. Sci., USA, 72
(1975) 3255.

(34) Ch. Beckers and H. Emmerich, Naturwiss., 60(1973) 50.

(35) R.C. Lauer, P.H. Solomon, K. Nakanishi, and B.F. Erlanger,
Experientia, 30(1974) 560.

(36) a. G.C. Levy and G.L. Nelson, Carbon-13 Nuclear Magnetic
Resonance for Organic Chemists, Wiley-Interscience, New
York, 1972; b. J.B. Stothers, in G.C. Levy (Editor),
Topics in Carbon-13 NMR Spectroscopy, Vol. 1, Wiley-
Interscience, New York, 1974, Ch. 6, 229.

(37) H. Hikino, T. Okuyama, C. Konno, and T. Takemoto, Chem.
Pharm. Bull., 23(1975) 125.

CHAPTER 7

ANALYSIS OF THE NATURALLY OCCURRING JUVENILE HORMONES - THEIR ISOLATION, IDENTIFICATION, AND TITER DETERMINATION AT PHYSIOLOGICAL LEVELS

DAVID A. SCHOOLEY

Biochemistry Department, Zoecon Corporation, Palo Alto, California 94304 (USA)

CONTENTS

INTRODUCTION

Juvenile hormones (JH) are secreted by the corpora allata of insects at specific intervals in the life cycle, and maintain the insect in an immature state until conditions favorable for maturation have been achieved. These hormones are also essential in many species for stimulating egg maturation and pheromone production in adult females.

In 1956 Williams [1] discovered an unusually rich source of juvenile hormone activity in the abdomens of male Cecropia moths. Several groups attempted the isolation and identification of the unstable hormonal principle contained in the "golden oil," or ether extract of cecropia abdomens. The first successful isolation of a pure juvenile hormone was reported in 1965 by Röller et al. [2,3, see also 4]. In 1967 Röller et al. published [5] the skeletal structure of the bis-homosesquiterpenoid epoxide **1** (JH I) as the juvenile hormone of *Hyalophora cecropia*, and shortly thereafter Dahm et al. reported [6] a chemical synthesis that allowed positive assignment of the stereochemistry as *trans,trans,cis* (2*E*,6*E*,10*cis*). However, the following year Meyer et al. [7] demonstrated the presence of a second hormone in Cecropia present in 1/4 to 1/7 the amount of JH I. These workers identified the new hormone as **2** (JH II), a lower homolog of JH I. The natural mixture of Cecropia juvenile hormones was shown to be optically active [8], and the synthetic unnatural isomer was found to have much lower biological activity [9]. In 1971 the absolute configuration of JH I was determined as 10*R*,11*S* by three groups [10,11,12].

It has been held that the corpora allata secretion is nonspecific with respect to both stage and order of insect as a result of many gland transplantation experiments [13]. Nevertheless, biological assay of synthetic samples of JH I and II and numerous structural analogs has shown a wide range in responses in insects of various orders [14]. This observation together with the finding of two discrete hormonally active species in one insect prompted many workers to attempt the isolation of juvenile hormones from other species. Röller and Dahm were able to demonstrate the presence of the same juvenile hormones I and II in *Hyalophora gloveri* [15] and *Samia cynthia*

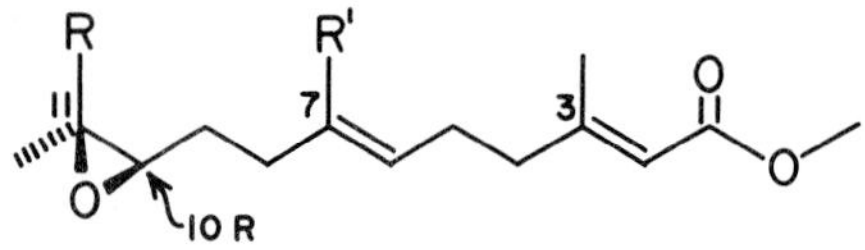

1 R = R' = Ethyl   JH I
2 R = Ethyl, R' = Methyl   JH II
3 R = R' = Methyl   JH III

1   methyl (2E,6E,10cis)-(10R,11S)-10,11-epoxy-3,11-dimethyl-
     7-ethyl-2,6-tridecadienoate [methyl 12,14-dihomojuvenate,
     C_{18} JH]

2   methyl (2E,6E,10cis)-(10R,11S)-10,11-epoxy-3,7,11-tri-
     methyl-2,6-tridecadienoate [methyl 12-homojuvenate, C_{17}
     JH]

3   methyl (2E,6E)-(10R)-10,11-epoxy-3,7,11-trimethyl-2,6-
     dodecadienoate [methyl juvenate, C_{16} JH].

Fig. 1. Structures and nomenclature of the three known insect
juvenile hormones 1-3. The IUPAC approved nomenclature is presented, according to which double bond geometry is designated
as E or Z, whereas ring structures (including epoxy) are designated as cis or $trans$. The popular trivial nomenclature JH I-
III corresponds to chronological order of discovery. Two less
commonly used trivial nomenclature conventions are indicated in
brackets.

[16]. These two silk moths are of the same family (Saturniidae)
as Cecropia, and share the peculiar ability of that species to
sequester large amounts of JH, a phenomenon that has very
recently been found [17] to result from an exclusive concentration of JH in the reproductive tract of adult male Cecropia
moths, specifically in the accessory gland. A number of groups
attempted to apply Röller's whole-body extraction techniques to
other insects with little or no success due to the much lower
content of juvenile hormones in insects of other families and
orders. An alternative approach to the juvenile hormone isolation problem was suggested by the discovery [18] that H.
$cecropia$ corpora allata can survive and produce juvenile hormon
when maintained in $vitro$. Following this lead, a research team
at Zoecon Corp. refined and simplified the gland culture
approach to the point where, in combination with other techniques, it was used to accumulate, isolate, and identify the

juvenile hormones of the Sphingid moth *Manduca sexta* [19]. The
in vitro results showed that adult female *Manduca sexta* CA produce two juvenile hormones, JH II and a previously unknown (in
Nature) sesquiterpenoid hormone **3** (JH III), but apparently
secrete none of the higher homolog JH I. The discovery of a
new juvenile hormone even within the order Lepidoptera
supported the concept of diversification of juvenile hormone
structure among the Insecta. Encouraged by the results of these
initial efforts, the group at Zoecon applied the same methodology to the isolation and identification of a juvenile hormone
of the grasshopper *Schistocerca vaga* (order Orthoptera), and
found that *Schistocerca* produces JH III [20]. Conspicuous
absence of JH I and JH II in gland cultures of this insect, the
first non-lepidopteran animal to be successfully investigated,
provided the impetus for additional studies of other insect
orders. Subsequent studies by a number of groups summarized in
Table I have resulted in the isolation and identification of
juvenile hormones from additional Orthopterans and Lepidop-
terans, and also from Dictyoptera, Coleoptera, and Hymenoptera.
Until recently, only JH III had been isolated from non-lepidop-
terous insects. Lanzrein *et al.* [21] isolated JH I, II, and
III from *Nauphoeta cinerea* nymphs, but almost exclusively JH
III from the adults of that cockroach species, leading these
workers to speculate that JH I and JH II play predominantly
morphogenetic roles while JH III might be the gonadotropic hormone.

The results of all published JH isolations to date (Table
I) have failed to demonstrate the existence of any juvenile
hormones other than those three already known. However, several
of these studies have not monitored fractionations by bioassay,
but have been strictly chemical identifications of the known
hormones. Two recent reports have claimed the existence in
insect extracts of biologically active chromatographic fractions that are apparently not identical with JH I, II, or III.
Blight and Wenham [29] report that extracts of *Schistocerca
gregaria* contain two biologically active fractions neither
identical to the known JHs nor to methyl farnesoate or the
corresponding olefinic "precursors" of JH I and II. Paguia *et
al.* [30] examined whole body extracts of *Attacus atlas* moths

TABLE I

IDENTIFICATION OF THE JUVENILE HORMONES OF VARIOUS INSECTS [EXCLUSIVE OF DATA FROM NEW GLC/ELECTRON CAPTURE DETECTION (ECD) METHODS]

Genus/Species	(Order)	Stage	Juvenile Hormone			Method	Reference
			JH I	JH II	JH III		
Hyalophora cecropia	(Lepidoptera)	ad. m	++	+	+	WB	5*,7*,22
"	(")	"	+	+		IV	18*
Hyalophora gloveri	(")	"	++	+		WB	15
Samia cynthia	(")	"	++	+		WB	16
Manduca sexta	(")	ad. f		+	+	IV	19*
Heliothis virescens	(")	ad. f	+	++		IV	23
Melolontha melolontha	(Coleoptera)	ad. m,f			++	WB	24
"	(")	ad. m,f			++	WB	22
Tenebrio molitor	(")	ad. f			++	IV	25*
"	(")	ad. m,f			++	WB	22
Leptinotarsa decemlineata		ad. f			++	WB	22
Apis mellifera	(Hymenoptera)	ad. f workers			++	WB	22
Schistocerca vaga	(Orthoptera)	ad. f			++	IV	20*
Gastrimargus africanus	(")	ad. f			++	IV	20
Schistocerca gregaria	(")	ad. m,f			++	WB	22
"	(")	larvae & ad. m,f			++	H	26*
Periplaneta americana	(Dictyoptera)	ad. f			++	IV	27*,28
Blatta orientalis	(")	ad. m,f			++	WB	22
Leucophaea maderae	(")	ad. m,f			++	WB	22
Nauphoeta cinerea	(")	ad. m,f			++	WB	22
"	(")	ad. f	trace	trace	+	H	21*
"	(")	larvae		+	+	H	21*

Abbreviations are as follows: adult (ad.), whole body extracts (WB), hemolymph extracts (H), *in vitro* techniques (IV), present (+), present as major component (++), male (m), female (f).
*Isolation of juvenile hormones in these studies was monitored by bioassay procedures.

(Saturniidae) and found a highly active fraction that co-
migrated with JH III on thin-layer chromatography (TLC), but
which did not appear to correspond to any of the known hormones
using the methods of Trautmann *et al.* [24].

The isolation and identification of juvenile hormones from
other species is still a field of considerable interest. Therefore methodology used in determining the structures of the known
juvenile hormones will be presented, and the relative merits of
various spectrometric methods will be critically reviewed.
Since physiological titers of these hormones are usually the
order of a few ng/g of tissue, separation of pure JH from chemically similar lipids can be a more challenging problem than
structure determination. Purification methods used in fractionating insect extracts will be discussed, with special
emphasis on high resolution liquid chromatography (HRLC) and
gas-liquid chromatography (GLC).

A field currently receiving even more attention than the
isolation of new juvenile hormones (as of April 1976) is the
fast-evolving area of physiological titer determinations of the
known JHs. A few initial reports have now been published
dealing with this challenging problem, and the significance of
these methods and probable future directions will be discussed.

BIOASSAY PROCEDURES

An all-important prerequisite in the isolation of the
known JHs was the availability of reliable, sensitive, and at
least semi-quantitative procedures to allow location of
hormonally active fractions in the various purification steps.
Juvenile hormone bioassay methods have been reviewed critically
by Staal [31] and Bjerke and Röller [32]. Currently the most
popular technique is the extraordinarily sensitive *Galleria
mellonella* pupal assay, after de Wilde *et al.* [33], in which
wax moth pupae are deliberately wounded on the mesothorax, and
test material applied to the $\sim$1 mm^2 wound in a paraffin oil-wax
mixture. A positive response is indicated by patches of pupal
cuticle over the wound after the adult molt. The *Galleria* unit
(GU) is defined as the amount of substance per test animal
giving a positive response in 50% of the pupae. Various values

for the specific biological activities of JHs in the *Galleria*
assay have been reported, most falling in the range of 2-5, 2-5,
and 60-500 picograms per GU for JH I, II, and III, respectively.
A less sensitive assay using *Tenebrio molitor* was utilized by
Röller's group in their structure elucidation of JH I. Far
greater differences in specific biological activity among the
three hormones are observed in this procedure: 1.2, 33, and
20,000 nanograms per *Tenebrio* unit for JH I, II, and III
respectively [34]. This disparity in relative activities
caused Röller's group to overlook JH II in *H. cecropia* extracts.

     This serves to emphasize a major problem in isolation of
new juvenile hormones: ideally extracts of a given insect
should be bioassayed on the same animal, but this is not
usually feasible due to difficulties in developing assays and
other factors. Wide variations in sensitivity of response to
the three known natural JHs and their analogs are extensively
documented [14], with such activity differences accentuating
the difficulties inherent in bioassaying the JH of one species
on another.

     Other pitfalls to be considered are the reliability or
specificity of JH bioassays. It is widely held that these
assays do not yield false positive responses when solvent
blanks or similar controls are tested. On the other hand,
Schmialek isolated the common natural metabolites farnesol and
farnesal, as JH active substances according to a *Tenebrio* bioassay, from yeast cells and *Tenebrio* feces [35] and *Samia
cynthia* bodies [36]. Despite the low specific biological
activity of these isoprenoids, several workers suggested that
they were identical with juvenile hormone. Subsequent chromatographic analyses showed these substances to be non-identical
with juvenile hormone [2,3,37]. JH activity has even been
detected in extracts of beef adrenal cortex [38]. Since certain
common natural metabolites elicit a JH-type bioassay response,
extreme caution must be exercised in equating positive assay
responses with the presence of a genuine juvenile hormone in
various animal extracts. Yet another complication is the
existence of unknown substances in crude extracts of insects
that inhibit the response of the test organisms to JH. Röller
and Bjerke [2] found that low temperature precipitation removed

inhibitory substances. Meyer *et al.* have discussed this
phenomenon [39], which they found to be mitigated by the base
washing step of their procedure. Blight and Wenham [29] have
added JH I to crude extracts of fifth instar *Schistocerca
gregaria*, and found lower than predicted *Galleria* biological
activity in these extracts <u>and</u> low temperature supernatants.

STRUCTURE AND STEREOCHEMISTRY OF THE THREE KNOWN JHS

Despite the existence of prior excellent reviews [40,41,
42], it is useful to summarize here the structure determination
of JH I, performed with only 200-300 µg of pure material. The
electron impact (EI) mass spectrum of JH I showed a weak molecular ion at m/e 294, indicating a tentative formula of $C_{18}H_{30}O_3$.
Since this formula demands unsaturation and/or cyclic struc-
ture(s), catalytic hydrogenation (Pd/C; C_2H_5OH) was performed
and the single product analyzed by mass spectrometry (MS). The
molecular ion indicated uptake of 3 moles of hydrogen and loss
of oxygen, and careful analysis of MS data showed the deoxo-
hexahydro derivative **4** to be a branched chain acyclic fatty acid

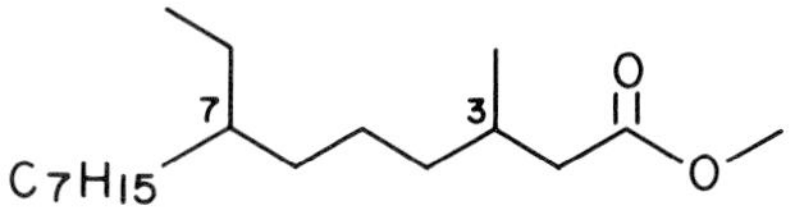

methyl ester. The fragmentation pattern clearly indicated a
methyl substituent on C-3 and an ethyl on C-7. The proton
nuclear magnetic resonance (NMR) data already in hand gave evidence for the existence of the following functionalities:

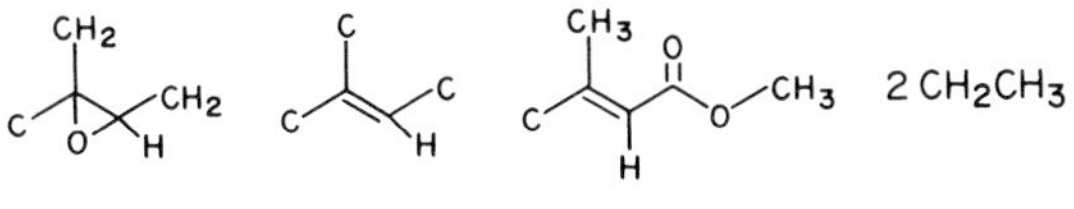

This allowed assignment of one site of unsaturation as 2*E*.
To locate the position of the additional unsaturation and the
suspected oxirane ring, 15 µg of hormone was degraded (osmium
tetroxide and periodic acid) to two dicarbonyl compounds, and
gas chromatographic comparison with a series of standards
revealed that the products appeared identical with levulin-

aldehyde (**5**) and, apparently, a one carbon homolog (**6**) of this

compound. Taken together with the previous data, this sufficed
to define the gross skeletal structure as **7**, with NMR data of
the C-6 vinyl proton indicating a <u>probable</u> 6*E* geometry.   Dahm
et al. [6] devised syntheses of isomers by routes in which

double bond geometry of precursors could be determined.   Subsequent NMR and gas chromatographic comparison of the various
isomers with natural material left no doubt that JH I possessed
the 2*E*,6*E*,10*cis* configuration.

Trost [41] states that this structure elucidation was
aided by studies of the model compound methyl 10,11-epoxy-*E*,*E*-
farnesoate (racemic **3**).   Hydrogenation of this compound (see
above) also revealed uptake of three moles of hydrogen and
expulsion of one oxygen--a most reassuring finding in light of
the usual resistance of epoxides to hydrogenolysis.   Also, the
MS data of the model compound clarified several fragmentations
in the mass spectrum of its 2-carbon homolog, "Cecropia JH," or
JH I.   This supplies a fine note of irony to the structure elu-
cidations of all three known JHs.   The model compound methyl 10,
11-epoxyfarnesoate was synthesized (together with various other
epoxyfarnesol derivatives) and bioassayed by Bowers *et al.* [43],
who found it to possess the highest known JH activity of any
"unnatural" substance tested up to that time on *Tenebrio* and
Periplaneta.   They speculated that "this compound will be found
to be very similar chemically to the natural [Cecropia] hormone
when the latter is isolated and characterized" [43].   Curiously
enough, eight years later the "model compound" itself was shown
to be a genuine secretory product of insect corpora allata [19,

20], albeit in optically active form.

Meyer, Schneiderman and their colleagues had also been
working on Cecropia JH for some years, and were aware of the
existence of two biologically active components in Cecropia
extracts, designated compounds B and E [7,39,44]. They
obtained additional spectral data (high resolution mass spectra
and infrared spectra) not recorded by Röller *et al.*, and concluded that their compound B indeed possessed the structure
proposed by the Wisconsin group for JH I. Their studies also
were aided by comparisons of the model compound methyl 10,11-
epoxyfarnesoate with compounds B and E, and a homology relationship of the three was indicated by MS. In addition, the
only significant differences in the NMR and MS data of these
three compounds were contributed by the functionalities at C-7
and C-11, fortuitously attached to the centers which had caused
the most "trouble" in the JH I identification. Accordingly
structure **2** could be assigned for compound E (JH II) entirely
from comparative spectral data without recourse to synthesis of
various isomers. A subsequent stereoselective synthesis of JH
II by Johnson *et al.* [45] did provide confirmation of the
assigned structure.

In 1970 Meyer and Hanzmann [8] subjected a 9:1 mixture of
natural Cecropia JH I and JH II to polarimetry measurements,
observing a positive dispersion curve with molecular rotation
at 589 nm ($[\alpha]_D$) of about +7°. These data did not allow definition of the absolute configuration of the two chiral centers
of the oxirane ring. The next year Loew and Johnson [9] and
Faulkner and Petersen [10] simultaneously published similar
syntheses of both JH I antipodes. The more stereoselective
route of Loew and Johnson gave (+)JH I that was <u>ca.</u> 9 times
more active on *Galleria*, and 6-8 times more active on *Tenebrio*,
than (-)JH I. Schooley and Nakanishi (footnote 3d in [9])
determined that Johnson's (-)JH I contained about 8% of the (+)
antipode, implying that the bioactivity of the unnatural antipode could be largely due to residual (+) hormone. Faulkner
and Petersen synthesized optically active hydroxy ketals **8** of
known absolute configuration, and the *S*-precursor led to (+)JH
I, allowing assignment of the absolute configuration of JH I as
10*R*,11*S* (since the oxirane ring is known to be *cis*, defining the

chirality of one center is sufficient to define both). Interpretation of these results was somewhat clouded by incomplete
separation of the *erythro* and *threo* diol intermediates in the
synthesis [10], giving samples of (+) and (−) JH I, each sub-
stantially contaminated with 2*E*,6*E*,10*trans* JH I geometric isomer
of opposite configuration at C-10.

   Within a few months this stereochemical assignment was
confirmed. Nakanishi *et al.* [11] and Meyer *et al.* [12] studied
the mechanism of acidic hydrolysis of synthetic JH II and I,
respectively, using $H_2^{18}O$. Mass spectral analysis of the 10,11-
diols by the two groups revealed that 97% [11] or 94% [12] of
the attack of $H_2^{18}O$ was at tertiary carbon C-11, thus implying
nearly complete retention of configuration at C-10 on hydro-
lytic opening.* Nakanishi's group next hydrated samples of

Johnson's (+) and (−) JH I and analyzed the optically active
diols by a then new method for determining absolute configuration of diols; circular dichroism measurements of the complex
formed in a solution of optically active diol and *tris*-
(dipivaloylmethanato)praseodymium. Comparison of the observed
Cotton effects of complexes from the antipodal diols with those

---

*Meyer *et al.* claimed a 5:2 ratio of inversion to retention at
C-11 (the site of nucleophilic attack of water), whereas
Nakanishi *et al.* observed essentially complete inversion at
C-11. This discrepancy may be due to slight differences in
the hydration conditions, but has no impact on the conclusions
reached by both groups on the configurations at C-10.

from four model diols of known absolute stereochemistry, established that JH I and its 10,11-diol possess the 10R configuration. Meyer *et al*. developed a scaled-down version of Horeau's
procedure [46] for partial resolution of α-phenylbutyric acid
by secondary alcohols. This method, applied to the 10,11-diol
from acidic hydration of natural *H. cecropia* (+) JH I, also
indicated a 10R configuration. Thus the independent results of
three groups were in agreement.

Following the discovery of Röller and Dahm [18] that
excised brain-corpora cardiaca-corpora allata complexes of
H. cecropia will survive and secrete JH *in vitro*, Judy *et al*.
at Zoecon refined the *in vitro* approach. They used a chemically
defined medium in which L-[*methyl*-^{14}C]methionine replaced the
non-labeled amino acid. Corpora cardiaca-corpora allata complexes from adult female *Manduca sexta* were maintained in this
system for periods up to several weeks. TLC analysis of an
ethyl acetate extract of the medium revealed a radioactive, biologically active zone approximating the R_f of the known JHs.
Further purification of this zone by high resolution liquid
chromatography (HRLC) revealed the presence of two labeled peaks
which also showed high biological activity. GLC/MS analysis of
the products of 738 gland pairs showed the more abundant (8.7
µg) hormone to be 85% pure and to have a mass spectrum nearly
identical to synthetic JH II, while the less abundant hormone
(5.3 µg) was 93% pure and had mass spectrum nearly identical to
methyl 10,11-epoxyfarnesoate. The only differences in MS data
between synthetic standards and natural hormones was attributable entirely to the remarkably high (71-73 atom %) carbon-14
content in the methyl group of the ester moieties, biosyn-
thetically derived from the 78 atom % L-[*methyl*-^{14}C]methionine.
Further proof of the structure and stereochemistry was obtained
microchemically. Aliquots (12-500 ng) of radioactive JH II or
III from culture were diluted with 50-100 µg of the appropriate
synthetic carrier. Cleavage of the 10,11-diols of each diluted
hormone with periodate followed by borohydride reduction gave
the same radioactive primary alcohol from each hormone, thus
suggesting a common skeleton to C-10 with 2*E*,6*E* geometry.
Radiolabel from diluted JH II diol eluted only with the threo
glycol (from the 10-*cis* epoxide), easily separable on HRLC from

the erythro isomer (from 10-*trans* epoxide impurity in the synthetic carrier). The 10,11-diols of both hormones were converted to esters of (+)-α-methoxy-α-trifluoromethylphenylacetic
acid (MTP esters). The diastereomeric esters formed from (+)-
acid and the two optical antipodes of each racemic carrier were
resolved on HRLC, but only the faster eluting peak in each case
bore radiolabel, thereby implying optical activity and optical
purity. The MTP ester of methyl 10,11-dihydroxyfarnesoate of
known 10*S* configuration was analyzed and found to be the
"unnatural" diastereomer, showing JH III to possess the same
10*R* absolute configuration as JH I. While the absolute configuration of JH II has never been as rigorously established as JH
I [10,11,12] and III [19], the faster eluting diastereomer from
(+)-MTP acid and racemic 10,11-diol of JH I, II, or III is in
each case the "natural" hormone derivative [19,23], thus indicating a 10*R*,11*S* configuration for *in vitro*-derived *Manduca*
[19] and *Heliothis* [23] JH II.

SPECTRAL TECHNIQUES FOR IDENTIFICATION OF JUVENILE HORMONES

Mass spectrometry (MS)

Mass spectrometry is undoubtedly the most useful tool for
structural verification of purified samples of juvenile hormones. Not only is a wealth of structural information derivable from MS, but even instruments manufactured in the 1960s
allowed determination of complete mass spectra with only microgram quantities. Coupled gas-liquid chromatography-mass spectrometry (GLC/MS) instrumentation has enabled mass spectral
determinations of samples that are still not totally pure. The
particular attractiveness of the GLC/MS technique is enhanced
further by the astounding sensitivity of newer units,
especially quadrupole-type spectrometers, which can determine
mass spectra with a few nanograms of sample. In view of the
difficulty of isolating even micrograms of JH, the advantages
of the method are obvious.

The electron impact (EI) mass spectra of JH I, II, and III
are given by Meyer [39,44]. The mechanisms of important fragmentations in the EI-MS of JH III have been investigated in
detail by Liedtke and Djerassi [47]. Both groups determined

precise elemental composition of important fragment ions by
high resolution MS. Since relative abundance of fragment ions
can depend significantly on experimental parameters such as
ionizing voltage, source temperature, and type of instrument
(especially magnetic sector *vs.* quadrupole), researchers
utilizing EI-MS for identification of the known JHs will doubtless prefer to compare spectra of unknowns with authentic
reference standard hormones run on the same instrument under
identical conditions. Consequently, there is little point in
reproducing plotted spectra in this work.

Chemical ionization (CI) mass spectrometry was utilized in
the identification of JH I, II, and III in *Nauphoeta cinerea*
[21]. Using isobutane as reagent gas, CI-MS revealed only three
principle ions from each JH, corresponding to the quasi-molecular ion MH^+, MH^+-H_2O, and MH^+-CH_3OH. The fragmentations are
due presumably to protonation and subsequent loss of the epoxy
and methoxy moieties, respectively. The high percentage of
total ion current carried by these peaks allows determination
of complete mass spectra with minute samples (according to [21],
4 ng of JH I injected into the GLC/MS was sufficient). However,
the lack of fragmentation supplies fewer diagnostic ions than
the corresponding EI mass spectra, and would seem to be of
lesser value in determining the structure of a hypothetical
"new" JH.

Fortunately, changing reagent gases in CI-MS alters the
observed degree of fragmentation, methane leading to more
extensive fragmentation than isobutane, and hydrogen more still.
Aprotic gases such as helium ionize by "charge exchange" rather
than proton transfer when used in a CI mode and lead to electron impact-like mass spectra. Methane CI-MS of JH III indeed
shows more fragment ions than the isobutane CI-MS [21].

The differences in EI and CI mass spectra of the juvenile
hormones provide extremely useful complementary information,
especially considering the virtual absence of molecular ions in
EI. In this regard, a technique that has shown considerable
promise in other applications is use of argon-water [48] or
helium-water as CI reactant gases, providing "hybrid" spectra
possessing strong quasi-molecular ions and the wealth of fragmentation seen in EI.

Other techniques

Proton NMR spectrometry is perhaps the most popular tool
of natural products chemists for deduction of structural, and
particularly stereochemical, information--since the latter is
usually not readily derivable from mass spectra. Unfortunately,
even the most sophisticated Fourier transform NMR spectrometers
currently require microgram quantities of scrupulously pure
material to produce useful information. Although Röller et al.
[5] had determined the proton NMR spectrum of purified JH I,
they still needed to rely on microchemical techniques to derive
the tentative chemical structure 1. In first identifying JH
III as an authentic product of insects, we did not obtain any
NMR data, but were able to elucidate all necessary stereochemical information, including even the absolute configuration, by
suitable microchemical techniques [19]. Microchemical degradations and reactions have long played a far more important role
than NMR spectroscopy in identifying microgram quantities of
insect pheromones (see Beroza [49]). It seems unlikely that
NMR will play a significant role in future juvenile hormone
research.

Meyer et al. [7,39] confirmed the presence of an α,β-
unsaturated ester functionality in JH I and JH II by infrared
spectroscopy. Judy et al. [19] utilized ultraviolet absorbence
(UV) spectra in identifying JH III and JH II from M. sexta
gland cultures. The UV data merely verified the presence of
the α,β-unsaturated ester moiety. Thus both techniques are of
rather limited value, but do supply added confirmation for
thorough studies. Measurement of UV spectra of the three JHs
requires very pure samples and solvents, since λ_{max} is 215 nm
in hexane, or 220 nm in methanol. In fact, Meyer [44] discussed the "misleading" information from UV spectral data of
partially purified JH, which showed only end absorption.

SPECIAL PRECAUTIONS CONCERNING JUVENILE HORMONE INSTABILITY

The juvenile hormones, as unsaturated epoxy esters, have
several functionalities capable of contributing to chemical
instability. The epoxide ring is susceptible to attack by a
variety of nucleophiles and is quite acid and heat labile.

Aqueous or alcoholic protic acids will quickly convert JH to
glycols or alkoxyhydrins, respectively. Lewis acids such as
boron trifluoride or stannic chloride promote cyclization of
racemic JH III to mixtures of various mono- and bicyclic
products [50]. Certain transition metals in aqueous solution
may well catalyze similar reactions [B. D. Hammock, personal
communication]. We have found that the acid catalyzed hydrolysis on a submicrogram scale frequently gives poor yields of
diols with formation of less polar materials suggestive of the
Lewis acid cyclization products [51]. Thermolysis of JH at
150-165° for a few hours in glass vessels affords excellent
yields of allylic alcohol(s) [52]. Anderson *et al.* [52] also

reported the use of neutral alumina (Woelm, Activity I) to
effect rearrangement of JH to allylic alcohols (as in thermolysis). Complete conversion of JH to allylic alcohols was noted
[R. J. Anderson, personal communication] in 24 hrs (38 mg of JH
in 1 ml hexane on 1 g Al_2O_3).

On the other hand, the ester function of JH is rather
stable towards base, probably due to the stabilizing effect of
conjugation. Attempts to separate JH from natural lipids by
selective saponification met with limited success [39].

Meyer has rightfully emphasized [39,44] the need to use
only carefully purified solvents for all work dealing with
small amounts of JH. He maintained that heating glassware to
200°C is a necessity of JH isolation work, and our experience
in developing JH analyses at nanogram levels bears this out (we
prefer 500°C). Use of chromic acid, or other acidic cleaning
solutions for glassware, is clearly contraindicated by the
chemistry of the epoxide group. Adsorption of JH to glass from
aqueous solutions is a problem, but can be mitigated by
treating glass with Carbowax 20M [53].

Samples of JH can be stored without appreciable decomposition even for years, if stored in a (dark) freezer at -20°C (or
lower), either: (a) neat, in a clean glass container such as a

conical flask or Kontes Reacti-Vial[TM], in which minimal surface
is exposed to the atmosphere, or (b) in Nanograde hexane
(preferably distilled) in a heat-treated glass-stoppered flask.
Since double bonds are generally prone to oxidation and other
degradative processes, storage of samples under inert gas is a
wise precaution.

A problem of quite another sort exists in laboratories
engaged in JH isolation or low-level analysis--that of the
manipulation or purification of milligram quantities of various
synthetic samples of JH used for chromatographic standards or
biological assays. Weighable samples of JH--some 2×10^8
Galleria units per mg of JH I or II--must be handled in such a
laboratory with extreme caution. Injection of even microgram
quantities of synthetic JH standards on silica HRLC columns,
despite the near quantitative recoveries observed in such systems, leads to a long term "bleed" of low levels of JH from the
chromatograph, rendering the system of dubious value for subsequent fractionation of picogram levels. Despite the
notorious thermolability of JH, similar contamination problems
were encountered in preparative GLC separations, in which
teflon parts from the heated collection port were found to
"bleed" JH activity [K. H. Dahm, personal communication]. Most
plastics have an avid affinity for JH and are to be avoided
insofar as possible.

PURIFICATION OF JH FROM EXTRACTS

The extraction and purification procedures for isolation
of JH I and II from Cecropia have been previously reviewed [39,
40,44]. While efficient, the purification steps were designed
for an animal containing unusually high JH titers, and were
also carried out before the widespread use of HRLC. It seems
useful to review critically the various techniques used to date
in fractionating whole body extracts to obtain JH, since it may
enable researchers to choose those techniques most suitable for
their particular problem. For example, hemolymph extracts can
be processed far more simply than whole body extracts due to
their lower lipid content.

Extraction techniques

The classic solvent for extraction of JH from *H. cecropia* [1,2,39] and other sources [24] is diethyl ether. After multiple extractions, filtration, and evaporation *in vacuo*, an oil is obtained which represents a high percentage of the wet weight of the organism (∿47% for *H. cecropia* [40]). However, de Wilde *et al.* [33] found that 6:1 ether-ethanol gives a much more rapid extraction of lipids and JH from water-diluted hemolymph (6 extractions necessary with alcohol-free ether). In his book on lipid analysis, Christie mentions the need for ethanol in removing lipid from lipoprotein fractions [54]. In developing alternative purification schemes for JH in connection with titer determination methods, we have utilized [51,55] methanol for extraction of both whole bodies and hemolymph. The filtered methanol extracts are not concentrated, but partitioned between pentane and 1-5% aqueous sodium chloride. On evaporation, the water washed pentane gives a lipid extract usually comprising only 1-2% of the wet sample weight of selected insect species.* The extraction-partitioning procedure was adapted from a general method [56] for analyzing many plant and animal tissues for the presence of traces of methoprene (trademarked Altosid by Zoecon Corporation), a JH analog used as a Dipteran larvicide.

Composition of crude insect extracts

The principal constituents of the crude lipid extracts of insects are glyceride esters of fatty acids. Other constituents are sterols (free and esterified), free fatty acids, waxes, cuticular hydrocarbons, and smaller quantities of many other non-polar materials. Polar lipids appear to be absent from the methanol/pentane/water extracts [D. A. Schooley and B. J. Bergot, unpublished], but might be present in ether extracts, although this does not seem to have been investigated. Marked differences in total lipid content exist between insect species.

---

*A notable exception was adult male *Samia cynthia*, in which ∿6.5% of wet weight appeared in the crude lipid extract; lipid from female *S. cynthia* was a more usual 1.9% of wet weight.

Considerable fluctuations are also observed in total lipid content within a species during the course of development. Moreover, distribution of total lipid between various structural classes depends on species, life stage, and tissue analyzed. Therefore the qualitative and quantitative importance of various lipid classes in crude insect extracts is somewhat unpredictable, complicating development of a generally useful scheme of JH isolation.

Low temperature precipitation

Röller and Bjerke [3] reported that dry ice cooling of the ethereal extract of *H. cecropia* reduced the mass of lipid in the filtrate by 3.8X. In studies on the related species *H. gloveri* [15], Dahm and Röller found that addition of an equal volume of methanol to the ether extract gave a low temperature filtrate with enhanced (10X) reduction in residual lipid. They subsequently reported that recovery studies with $[2-^{14}C]JH$ I revealed a 30% loss of JH with the latter technique [16]. Trautmann *et al.* used the latter method [15] with ether extracts of *Melolontha melolontha* [24], reporting a 5.3X reduction in biomass. The low temperature presumably causes crystallization of high melting sterols and saturated fats, leaving more unsaturated components in the filtrate [39].

Molecular distillation

Theory and practice of molecular distillation is reviewed by Vogel [57]; briefly, use of a short path between reservoir and condenser and very high vacuum (at least 0.001 Torr) results in the mean free path of molecules present being greater than the dimensions of the apparatus. This constitutes a "perfect" vacuum for the sample, and relatively large and unstable molecules can be distilled. In such an apparatus maintained at 60-100°, JH is volatile, whereas most higher molecular weight lipids (di- and triglycerides, sterols and their esters) are not. Meyer *et al.* [39] used a wiped-film apparatus for molecular distillation of crude lipid extracts of *H. cecropia*, achieving a one-step reduction in biomass of 19-23X, depending on batch of material [44]. Röller *et al.* [4] used molecular

distillation (short-path apparatus) on the concentrated filtrate from the low-temperature precipitation step and found an
8.7 reduction in biomass; the lower enhancement of purity is
likely attributable to removal of some saturated fats in the
prior step. Meyer *et al.* [39] report that the distillate from
H. cecropia extract consists of 92% free fatty acids, allowing
a subsequent facile, selective purification by simple partitioning between ether and aqueous base.

     Clearly this is one of the more efficient procedures for
removal of high molecular weight lipids from JH, especially for
very large-scale experiments [39]. Although use of labeled JH
to determine precise recovery was not reported, Röller *et al.*
[4] claim little discernible loss in bioactivity.

Column and thin layer chromatography

     Virtually all JH isolations to date have utilized a silica
or alumina chromatography for purification at some point in each
scheme. Since these adsorbents are disposable and rather cheap,
their use is indicated at a relatively early stage of purification. Generally extracts should not be subjected to HRLC or
GLC purification and/or analysis without a preliminary clean-up
by adsorption chromatography (TLC or column), since the useful
life of expensive HRLC and GLC columns can be appreciably
shortened by injection of non-prepurified extracts. Use of pre-
washed and deactivated adsorbents is desirable, particularly
considering the previously mentioned instability of JH on
alumina.

     Comparison of relative purification efficiencies of
various adsorption chromatographic systems utilized in the
literature is difficult due to diversity of sample composition,
except for a number of experiments with crude Cecropia oil. If
a crude lipid extract is utilized, the problem can be largely
one of separating JH from tri- and diglycerides. Methods
developed for adult male *H. cecropia* may not be directly applicable to whole body extracts of other insects without modification. The data in Table 2 summarize those adsorption chromatography systems described in the literature for which purification factors are quoted. The method apparently tested with
the widest variety of insect species is the last listed in

Table 2. Of the two solvent mixtures quoted, 4% tetrahydrofuran in dichloromethane causes JH to migrate equidistant
between di- and triglycerides, whereas with 5% butanone in
dichloromethane, JH is closer to diglycerides. In either case,
cholesterol is more polar than JH or diglyceride [B. J. Bergot
and D. A. Schooley, unpublished observations].

Gel chromatography

Sephadex LH-20 eluted with 1:1 benzene-methanol, was utilized by Dahm and Röller [15] for purifying low-temperature
filtrates of *H. gloveri* crude extracts, a 14X purification
being achieved. Bieber [59] utilized the same system and determined that JH apparently elutes with the leading edge of
the sterol peak. More recently, Dahm *et al.* [34] substituted
acetone for methanol and found 80-100X purification of JH from
H. cecropia crude oil (LH-20 column eluted with 1:1 benzene-
acetone). Sephadex LH-20 tends to adsorb the polar component
of the solvent mixture in such systems and separates predominantly by a partition mechanism [60]. Trautmann *et al.*
[24] purified extracts of *Melolontha melolontha* by similar
methods but substituted neat chloroform as eluent and achieved
a 12X purification. The separation mechanism of Sephadex LH-20
with chloroform is thought [60] to be based on molecular size,
or gel filtration chromatography.*

Classic gel filtration chromatography, or use of various
Sephadex supports eluted with aqueous buffers, has been
utilized in several studies of JH biochemistry. Kramer *et al.*
[61] determined the aqueous solubility of JH I (5×10^{-5} M in
0.02 M Tris buffer, pH 7.3), and stated that the elution profile of this solution on Sephadex G-15 showed the absence of
high molecular weight aggregates of JH. Pratt and Tobe [62]
used thin layer gel filtration (Sephadex G-75 superfine) to
demonstrate that JH is released by locust corpora allata as a
free solution rather than as a protein complex. B. D. Hammock
[personal communication] observed useful fractionations of

---

*Modern usage prefers the term gel permeation chromatography
when organic solvents are utilized.

TABLE 2

CHROMATOGRAPHIC PURIFICATION OF JUVENILE HORMONES FROM LIPID
EXTRACTS OF *H. CECROPIA* AND CERTAIN OTHER INSECTS

Adsorbent	Technique	Solvent	Purification Factor	Type Extract	Ref.
Silica-active	CC	∿1% Ethyl Acetate in Benzene	10	C	58
Silica-8% H_2O	CC	85% Benzene in Cyclo-hexane	100	C	39
" "	CC	"	18	PP	39
Zinc Carbonate -1% H_2O *	CC	∿10% Benzene in Hexane	35-60	PP	39, 58
Alumina-4.3% H_2O	CC	20-36% Benzene in Heptane	75	C	58
"	CC	"less abrupt increase in Benzene con-centration"	75x3	C	58
Alumina	CC	CCl_4 in Cyclo-hexane	8.6	PP	44
Silica	TLC	$CHCl_3$-Ethyl Acetate (2:1)	12	PP	4,40
Silica	TLC	$CHCl_3$:Pentane (2:1)	8	PP	4,40
Florisil**	CC	2% Ethyl Acetate in hexane	12	PP	24
Silica***	TLC	4% THF in CH_2Cl_2 or 5% Butanone in CH_2Cl_2	20-140 (usually ∿50)	C	55

Abbreviations are as follows: thin layer chromatography (TLC),
column chromatography (CC), crude extract (C), pre-purified
extract (PP), distilled tetrahydrofuran (THF).

* Subsequently found to be non-reproducible and to cause
  excessive loss of hormonal activity [39].

** Extracts of *Melolontha melolontha*.

***Extracts of *Aedes aegypti, Tenebrio molitor, Manduca sexta,
   Diatraea grandiosella, Sarcophaga bullata, Samia cynthia.*

insect hemolymph on Sephadex G-150, and suggested that the
lipid-free fraction containing JH bound to its specific carrier
protein could serve as the basis of a highly selective and
efficient purification. A similar system has found use in
studies by Nowock *et al.* [63] of the synthesis of JH binding
protein. In handling aqueous solutions of JH, the previously
mentioned precautions regarding adsorption to glassware must
be borne in mind. Sanburg *et al.* have shown that unbound JH
in aqueous solution is rapidly and completely adsorbed by
charcoal [64].

Partitioning techniques

In early work on isolation of JH from *H. cecropia*, Gilbert
and Schneiderman [65] purified the crude oil using a 50-trans-
fer countercurrent distribution (CCD) with 1:1 petroleum ether
(b.p. 30-60°)-83% aqueous ethanol, and found partition coefficient K = 2.3 (favoring the hydrocarbon phase). A 50X
purification was achieved in one pass, but only an additional
2X in a second pass. In attempting to improve this system,
Meyer and Ax [66] sought to obtain a partition coefficient
near unity. Hexane-93% aqueous ethanol showed K = 1.0 at
18.5°C, but only 8.2X purification of a "washed" Cecropia oil
was observed in a 300-transfer CCD.

Meyer and Ax [66] also reported a system for reversed-
phase partition chromatography of JH. Whatman 3MM paper was
dipped in a 15% heptane solution of Bayol D (Esso, a kerosene-
type white oil), air-dried, spotted with Cecropia oil, and
eluted with 40% aqueous propanol. The JH-active zone had
R_f=0.24. The method is quite slow and has not been widely
utilized in subsequent work.

Simple solvent partitioning procedures using separatory
funnels have shown some utility in JH isolations. Meyer *et al.*
[39,44] had partitioned crude Cecropia oil between cyclohexane
and 85% aqueous methanol in their first attempt at large-scale
isolation of JH and obtained an impressive 13X enhancement in
purity, but did not utilize the method subsequently.
Trautmann *et al.* [24] used an exhaustive solvent partitioning
to purify large residues from low temperature filtration of a
methanol-ether solution of *M. melolontha* crude extract. A

hexane solution of residue was washed nine times with acetonitrile, and the combined acetonitrile extracts evaporated, redissolved in hexane, and the hexane again washed eight times
with acetonitrile. Concentration of the acetonitrile gave a
4.2X purified JH residue. In our scheme for extracting and
purifying JH from whole bodies for GLC/ECD analysis [51,55],
part of the apparent efficiency of the extraction procedure is
doubtless due to the subsequent partitioning step.

HORMONES PRODUCED *IN VITRO VS. IN VIVO*

Extraction of *H. cecropia* abdomens, an unusually rich
source of JH, yields a lipoidal oil which still requires five
purification procedures to produce reasonably pure hormones [4,
39]. In contrast, the *in vitro* method as refined by Judy *et
al.* [19] provides extracts that generally require only two
chromatographic steps, on a small and convenient scale, to isolate pure hormones. It has also been asserted [19] that the
amount of hormone available from each animal *in vitro* is
greater, due to continuous production and lack of degradative
enzymes, and that the system provides a practical source of
radiolabeled, optically active natural hormone(s) in aqueous
solution for certain *in vitro* studies on target organs. The
in vitro system is certainly ideally suited for biosynthetic
studies, since labeled precursors can be administered to the
corpora allata without concern for the metabolism of precursors
in other tissues. This technique served us to great advantage
in demonstrating the origin of the "extra" carbon atom in JH
II from propionate [67] *via* the possible intermediacy of homo-
mevalonate [68] (see Schooley *et al.* for a review of JH biosynthetic studies [51]). The failure of mevalonate to incorporate
into JH I of Cecropia *in vivo* [34] is in marked contrast to its
incorporation *in vitro* into JH II and III of *M. sexta* [67].
This discrepancy is possibly rationalized by the results of
Goodfellow *et al.* [69], who report a highly efficient and
unusual incorporation of mevalonate into triglycerides in live
Sarcophaga bullata.
    Critics of the *in vitro* system claim that the performance
of the glands *in vivo vs. in vitro* is likely quite different.

However, corpora allata of both *Tenebrio molitor* [25] and
Periplaneta americana [27,28] (adult females) secrete only JH
III when maintained *in vitro*, a finding fully confirmed by
analysis of whole body extracts (mixed sexes) of both species
[22]. Recently developed GLC/ECD assays for the known JHs [51,
70] indicate that adult female *M. sexta* possess roughly equal
hemolymph titers of JH II and JH III, together with smaller
amounts of JH I--a result admittedly somewhat different from
the observation that corpora allata from adult females of this
species maintained *in vitro* secrete comparable amounts of JH
II and JH III only [19]. Perhaps the greatest failing of the
in vitro method is the inability of corpora allata from
various insects, or various stages of the same insect, to produce JH. For example, Judy *et al.* found that adult male and
early fourth instar larval *M. sexta* corpora allata fail to
secrete JH *in vitro* [19], yet chemical assays [51,70] show
higher hemolymph JH titers in these larvae than in adult
female *M. sexta*, whose corpora allata secrete JH *in vitro*.

Recent advances in separation and spectral techniques may
make *in vivo* techniques more advantageous for hormone isolation from many insects. Nevertheless, the *in vivo* and *in
vitro* methods should best be viewed as complementary rather
than competing techniques, and we are fortunate to have both
methodologies available.

GAS-LIQUID CHROMATOGRAPHY AND HIGH RESOLUTION LIQUID CHROMATOGRAPHY

Both GLC and HRLC can be employed not only for micropreparative isolation, but are capable of quantitative analysis
with appropriate detectors, and offer greater resolution than
previously discussed separation methods. Dunham *et al.* [71]
recently reviewed the application of these techniques to JH
analysis.

Gas-liquid chromatography
Research leading to the structure determination of the two
juvenile hormones of *H. cecropia* relied heavily on GLC techniques. Meyer [44] has discussed extensively the construction

of a metal-free system for preparative separation of JH without
decomposition since commercial all-glass chromatographic systems were not yet common in the early-mid 1960s. A chemical
rationale for the problems encountered by Meyer is the facile
thermal rearrangement of JHs to allylic alcohols [52], a process possibly catalyzed by metal surfaces. Meyer *et al.* [39]
extracted their stationary phases with base in order to remove
acidic impurities that contribute to on-column decomposition of
JH. Dunham [71] recommends the use of a slightly basic
stationary phase, phenyldiethanolaminesuccinate (PDEAS), which
not only allows chromatography of these epoxides without decomposition, but also shows good selectivity in separating olefin
isomers.

Despite potentiality for substrate decomposition, the isolation scheme developed by Röller and Bjerke [3] and Röller *et
al.* [2,4] used GLC as the final isolation step to good advantage: pure material was obtained for the subsequent structure
proof [5] with over 300X enhancement of purity [2,3] and up to
80% recovery [2] - an extremely favorable result for a technique
notorious for poor recoveries. Figure 2 shows the preparative
isolation of 16 μg of JH I, after final TLC purification [4].
Separation of synthetic 10,11-epoxy *cis* and *trans* isomers of
JH I and II is only partial on packed columns. Trautman *et al.*
utilized 20 meter glass capillary (0.32 mm i.d.) columns with
Grob [72] injectors for GLC/MS analysis of JH from many insects
[22,24]. With capillary columns, all geometric isomers of
synthetic JH I are apparently separated [24].

Argon ionization detectors were used by Röller *et al.*
[2,3,4] and Meyer *et al.* [7,39] in isolating JH I and II. In
fact, Meyer and Knapp [73] devoted considerable effort to
improving existing detectors of this type, and claim a 1 ng
detection sensitivity. Despite the purported advantages of
high sensitivity and simple construction, injection solvents
can cause discharge currents harmful to the source [73], and
these detectors are neither widely used nor commercially
available. Modern flame ionization detectors (FID) serve quite
adequately for the same purpose and can detect as little as
1 ng under favorable conditions.

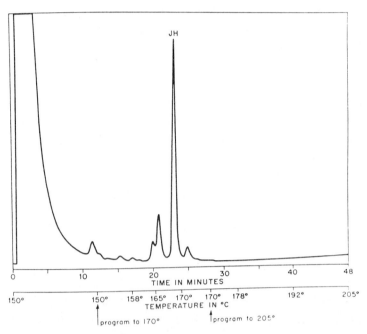

Fig. 2. Micro-preparative gas-liquid chromatogram of JH I from
H. cecropia. Whole body extracts were purified in four steps
prior to this final purification [4]. Conditions of analysis
were: 180 x 0.3 cm glass column, 3% XE-60 on 80-100 mesh Gas
Chrom Q, argon carrier at 75 ml/min, argon ionization detector
at 210°. Column temperature was isothermal at 150° from 0-12
minutes, then linearly programmed as indicated. Reprinted by
permission of Prof. H. Röller and Pergamon Press [from ref. 4],
(*J. Insect Physiol.*).

High resolution liquid chromatography (HRLC)*

HRLC has almost totally displaced GLC for micropreparative
isolations of JH because of the stability of the hormones on
HRLC, and the near quantitative recoveries routinely achieved
with this technique. Most separations of JH have been
performed with silica-based HRLC columns.

---

*This technique has also been termed HPLC (high pressure, or
high performance liquid chromatography) and HSLC (high speed
liquid chromatography). Since resolution is usually the most
important feature, rather than high back pressure or speed of
analysis, the term HRLC seems best to us and others [74].

Separation of the juvenile hormones and their derivatives
by HRLC was first accomplished in the determination of the
absolute configuration of JH I [11,75] and in the first isolation of JH III from *M. sexta* [19] (see Figure 3).

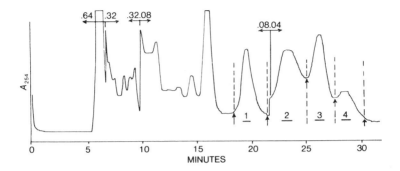

Fig. 3. Micro-preparative liquid chromatographic separation
of JH II and JH III from *Manduca sexta*. Extracts of culture
medium were purified by a single TLC step prior to HRLC.
Material in zone 2 was 85% pure JH II, while material from
zone 4 was 93% pure JH III, the first identification of this
substance in insects. Conditions of analysis were: 300 x
0.24 cm Corasil II, eluted with 6% diethyl ether in pentane at
1.0 ml/min. Changes in attenuation of the UV detector are
noted by double-headed arrows beneath numbers giving the range
in absorbance units full scale. Reprinted from *Proc. Natl.
Acad. Sci. USA*, [19].

These earlier separations of JHs and their derivatives by HRLC
used columns packed with Corasil II, a "pellicular" support
consisting of a glass bead core (~37-50 μm) thinly coated with
silica. Pellicular LC packings give generally higher efficiencies than comparably sized porous silica. More recently,
advances in column packing technology have allowed preparation
of short (0.15-0.5 m) columns of carefully sized spherical or
irregular silica in the 5 and 10 μm particle diameter range.
Such columns show as efficient mass transfer properties* as the

---

*Stationary phase mass transfer is a function related to the
residence time of the solute in the adsorbent particles, a time
governed by adsorption processes and diffusion into and out of
the particle. The diffusion factor causes the advantageous
nature of smaller particles and pellicular materials.

pellicular materials, but possess the advantage of considerably
increased capacity for solute mass, since they lack the inert
glass core characteristic of pellicular supports. We find far
better resolution of the JHs on a 22 cm Zorbax-SIL column
(Dupont Instruments, see Figure 4) than on a 300 cm Corasil II
column (cf. Figure 3).

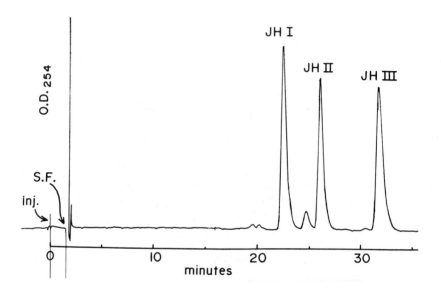

Fig. 4. High-resolution liquid chromatogram of synthetic
standards of the three juvenile hormones on a microparticulate
silica column. Impurity eluting between JH I and JH II is the
2E,6E,10trans synthetic isomer of JH II. Baseline resolution
of this isomer is not achieved by GLC using packed columns.
Conditions were: 22 x 0.46 cm Zorbax-SIL eluted with 6% diethyl
ether in pentane (50% water saturated) at 2.0 ml/min, attenuation 0.04 AUFS, ∿40 µg of each hormone, inj. = injection
point, S.F. = solvent front. Efficiency of column is ∿13-14,000
theoretical plates for JH I-III.

Two very important considerations in silica HRLC are the
water content of the silica and the role of the mobile phase,
or eluting solvent. For an excellent general discussion of
these variables see Snyder and Kirkland [76]. It is desirable
to maintain about half of a monolayer of water over the silica

active sites. Strong, water miscible solvents elute this bound
water rapidly, and even water immiscible solvents elute it,
although more slowly. In either case the eventual result is an
increase in activity and retentivity of the column, generally
coupled with lower efficiencies and peaks that tail badly. The
ideal situation is use of solvent containing enough water so
that a stable "half-monolayer" is achieved. Determining the
precise amount of water required is difficult; the usual
expedient is the use of solvents that are 50% saturated with
water [76]. While non-aqueous hydroxylic modifiers are doubtless easier to use (0.1-0.2% 2-propanol added to the HRLC solvent mixture), literature reports show that their use can lead
on occasion to abnormal, confusing peak shapes [76,77], and our
experience bears this out.

Selection of a proper HRLC solvent mixture must be considered critically [76]. The solvent(s) must be highly pure
(preferably distilled in glass), compatible with the detection
device, stable, and preferably readily volatile and non-viscous.
The frequent use of 254-nm ultraviolet absorbance detectors prevents use of many useful aromatic and ketonic solvents. Ethers
and alcohol-free chloroform must be purified just before use,
or avoided. Column efficiency is lower with more viscous solvents, so in choosing between the isopolar pair of hexane *vs.*
cyclohexane, the 3-fold lower viscosity of the former suggests
its use. We use *n*-pentane as the saturated hydrocarbon of
choice, since its volatility allows easy sample recovery and it
is extremely non-viscous: a cheap, commercial grade (Phillips
Petroleum) is free of the aromatics usually found in hexane.
For separation of JHs on silica, addition of 4-8% of diethyl
ether - or somewhat less ethyl acetate - gives suitable retention volumes. Occasionally, recourse must be made to other
solvent systems. For example, in isolating JH III from
Tenebrio molitor corpora allata *in vitro* [25], TLC-purified
extracts of culture media were subjected to HRLC purification
using 8% ether in pentane, but the JH III was only slightly
greater than 50% pure by GLC. Re-injection using neat dichloromethane as the HRLC solvent removed a major impurity,
yielding 83% pure JH III.

Alumina HRLC has been utilized in our scheme for purifying
"JH-PFPA" derivatives for subsequent analysis by GLC/ECD [51,
55]. Controlling water content of alumina HRLC is far more
difficult than is the case with silica, and in this application
we utilized 0.065% isopropanol as modifier [55]. Reversed-
phase chromatography of JH on chemically bonded hydrocarbon
packings, such as μBondapak C18 (Waters Associates), is a
highly promising technique. The order of elution of the three
hormones is the same as GLC, and the reverse of normal phase
(silica) HRLC. In an appropriate eluting solvent such as 70-
75% methanol in water, the chromophore experiences an appreciable "red" shift, greatly enhancing sensitivity of detection.
Since polar impurities elute first in reversed phase, there is
less danger of damaging columns with polar "pollutants" than
in silica HRLC. The major disadvantage of reversed-phase HRLC,
sample recovery from water-methanol mixtures, can be easily
solved by diluting the eluent with water and extracting JH with
pentane.

The standard 254-nm UV detectors have only moderate sensitivity for detecting JH in normal phase HRLC, since the absorp-
tivity at 254 nm is 5% or less of that at λ_{max} (∿215 nm in
hexane). Lanzrein et $al.$ [21] used a variable wavelength UV
detector set at 215 nm, and were able to detect as little as
50 ng of JH using 60 x 0.4 cm silica columns. Presently, such
detectors have poorer signal-to-noise ratio at 254 nm than high-
quality fixed-wavelength detectors. I find that using
reversed-phase chromatography (μBondapak C18, 30 x 0.4 cm) with
a 254 nm detector (Spectra-Physics Model 230), detection limits
are just as good as those quoted above [21] because of the
chromophoric shift and the superior signal-to-noise characteristics of this cheaper, simpler unit.

DETERMINATION OF PHYSIOLOGICAL TITERS OF THE THREE KNOWN
JUVENILE HORMONES

Determinations of the hemolymph JH titer in various insect
species has played a crucial role in contributing to knowledge
of the function of JH in developmental physiology. To date
titer determinations have relied on bioassay. While these

studies have advanced our knowledge of such processes as the
timing of JH secretion during development, investigators have
long awaited other methods that circumvent the limitations of
bioassay: imprecision and lack of discrimination between the
three known species of JH. Thus all bioassay titer studies of
insect tissue extracts provide only rough, cumulative titers
of the three known JHs (plus biologically active intermediary
metabolites!). The necessity for qualitative discrimination
between the known JHs seems to mandate a chromatographic separation of the hormones or their derivatives prior to, or concomitant with, detection. Therefore, it is appropriate to use
mode of detection as the basis for discussing various qualitative and quantitative analyses of JH. An ideal assay system
should be simple, rapid, specific, and precise, and utilize
readily available apparatus. The assay system ·hould be highly
sensitive, requiring less than 10 gram samples of insect
tissue, and be applicable not only to relatively "clean" hemolymph, but also to "dirty" whole body extracts for the many
cases in which hemolymph collection is difficult.

All schemes developed as of this date achieve several of
these goals, but at the same time fail in other aspects.
Nevertheless, the urgent need for such assays is bound to stimulate improvements in methodology. With the techniques at
hand, we are already beginning to see tentative answers to
questions such as: Why are there three juvenile hormones, and
do they play different physiological roles? Is there a phylogenetic variation in JH identity?

Gas-liquid chromatography-mass spectrometry (GLC/MS)

A procedure published in 1972 by Bieber *et al.* [59,78]
described a method of analysis for JH I (only) using GLC with
multiple ion detection (MID). [*Methoxy*-^{2}H$_3$]JH I was added to
Cecropia extracts as both carrier and internal standard, and
re-isolated by a method similar to that of Dahm and Röller
[15]. In the GLC/MID technique, the mass spectrometer serves
as no more than an expensive, highly specific detector for the
GLC: rather than scanning total spectra, the instrument con-
stantly monitors one or several fixed ions. Sensitivity is
greatly enhanced over that obtainable when full spectra of

several hundred amu are being scanned. Using the EI mode,
Bieber *et al.* focussed on m/e 114 (from natural, "protio"
hormone) and m/e 117 (from [*methoxy*-$^2H_3$] hormone), fragment
ions which are characteristic electron impact rearrangement
products of the unsaturated ester moiety. Certain higher mass

R=H, m/e 114
R=D, m/e 117

ions were also used (m/e 209 "protio," m/e 212 trideuterio),
but were considerably less abundant. The sensitivity of the
method was said to be 200 ng when 75 µg of carrier was used.
The method has apparently not seen subsequent use.

In 1974 Trautmann *et al.* published [24] the first general
method for qualitative and quantitative analysis for all three
JHs, and subsequently applied their method to studies of insect
species from four orders [22]. Preparation of samples for
analysis involved ether extraction of whole insect bodies and
addition of a trace of [^{3}H]JH I to monitor recovery, followed
by six purification steps. Good recovery of radiolabel was
observed (67%) considering the number of procedural manipula-
tions. In the final chromatographic purifications, JH I and
II were co-isolated, but JH III was separately isolated. Final
qualitative and quantitative analysis was achieved by injecting
onto glass capillary GLC columns using the Grob [72] method;
the coupled mass spectrometer gave highly specific identification of peaks presumed to be JH (major diagnostic fragmentations were observed, in one case even the molecular ion was
observed). Since a small quantity of JH I was added, data had
to be corrected for the exogenous hormone. In all species
investigated by Trautmann *et al.* [22,24], only JH III could be
identified. The method has been applied subsequently to analysis of the JH-active principle of *Attacus atlas* moths. While
this principle co-migrated with JH III on TLC, GLC/MS failed to
reveal the presence of JH I-III, implying the existence of a
new hormone [29].

Although this method has yielded valuable data, the sensitivity is rather low (100 ng). Therefore to observe the
desired limit of detection of 0.1 ng JH/g of tissue, a full
kilogram of insects must be extracted.

Lanzrein *et al.* [22] have isolated juvenile hormones from
hemolymph of the roach *Nauphoeta cinerea*. Their methodology
represents more accurately an isolation and identification
scheme rather than a general method of physiological titer
determination, but the sensitivity of detection merits its discussion here. Hemolymph from roaches was collected and
extracted in ether-ethanol (5:1) and purified by TLC. The JH
zone was separated on HRLC, and individual fractions corresponding to each JH were submitted to GLC/MS analysis on a quadrupole mass spectrometer (Finnigan 3300) in the CI mode.
Sensitivity for JH I was 4 ng injected on-column. The HRLC
fractions were independently bioassayed on *Galleria*, and relative quantification of titers of the three JHs was based only
on bioassay results. These workers found *N. cinereae* penultimate instar larvae to contain JH I, II, and III in approximate 2:1:3 ratio, while adults contained almost exclusively
JH III. This led to the interesting speculation that JH III
is predominantly gonadotropic in function, whereas JH I and II
are primarily morphogenetic hormones [21].

Work has been in progress for some time at Zoecon on a
GLC/MID procedure [D. A. Schooley, L. L. Dunham, and B. J.
Bergot, unpublished] for assay of the three JHs. This scheme
utilizes extraction and purifications in common with the Zoecon
GLC/ECD procedure [51,55], but it is shorter and far more
specific, albeit slightly less sensitive. In the author's
opinion such a GLC/MID assay may come closest to being the
ideal JH assay. Its only shortcoming is the need for expensive
instrumentation, a quadrupole GLC/MS/data system.

Gas-liquid chromatography with electron capture detection (GLC/
ECD)

The electron capture detector employs a radioactive foil
to ionize carrier gas, giving rise to a "standing current" when
voltage is applied between foil and another electrode. Certain
functionalities, especially organohalides and some unsaturated

systems, cause a decrease in standing current on passing
through the EC cell. The detector is thus selective and
remarkably sensitive; sub-picogram quantities of many halo-
genated materials are detectable.

Unfortunately, JH elicits little if any electron capturing
response, and use of the technique requires derivatization
schemes to introduce "electrophoric" functionality into the JH
molecule. The feasibility of this approach was first demonstrated by Dunham (see ref. 19), who subjected JH to acid-
catalyzed hydration and converted the resultant 10,11-diol to
the *bis*-trifluoroacetate. Application of this method to hemolymph extracts of early 4th instar larvae of *M. sexta* revealed
the probable presence of JH III in hemolymph, but results with
JH I and II were inconclusive [19]. A modification of this
procedure was published recently by van Broekhoven *et al.* [79],
who substituted heptafluorobutyric anhydride for trifluoro-
acetic anhydride, thereby obtaining the diol *bis*-heptafluoro-
butyrate. These workers analyzed hemolymph of *Leptinotarsa
decemlineata* for the presence of JH III only, claiming a value
of 17 ng/g. Experiences with the epoxide hydration reaction in
our laboratory [51] and elsewhere have been highly erratic.
Using conditions apparently identical to those of others [79],
we frequently find inexplicably low yields of diol (as low as
5-10%) from nanograms of epoxide, accompanied by a host of un-
related products. Such experiences caused Hammock *et al.* [53]
to develop an enzymic method for conversion of JH to the diol
derivative. Using partially purified, soluble epoxide hydra-
tase from mouse liver or kidney, they converted [^{3}H]JH I to its
diol derivative in high yield on a nanogram scale. It is pru-
dent or necessary in any case to add to extracts an internal
standard capable of monitoring recoveries in purification, and
yields in derivatization procedures,such as the *n*-propyl ester
of [^{3}H]JH I [51,55], prepared by transesterification of
commercially available [^{3}H]JH I (New England Nuclear).

In contrast to the epoxide hydration, acid catalyzed
methanolysis of the JHs to yield 11-methoxy-10-hydroxy derivatives ("methoxyhydrins"), generally proceeds well on a nanogram scale [51]. Two groups of researchers have recently
developed GLC/ECD methods based on acylation of the JH methoxy-

hydrin derivatives with "electrophoric" haloaromatic acids. Publications describing the methods and certain biological applications are just available [55] or still "in press" [51, 34,70].

An application of the Zoecon method is described in a collaborative study [55] of JH titers in *Diatraea grandiosella* with Chippendale and Yin of the University of Missouri. An outline of procedural steps for processing hemolymph or whole body extracts is outlined in Figure 5. Between the initial

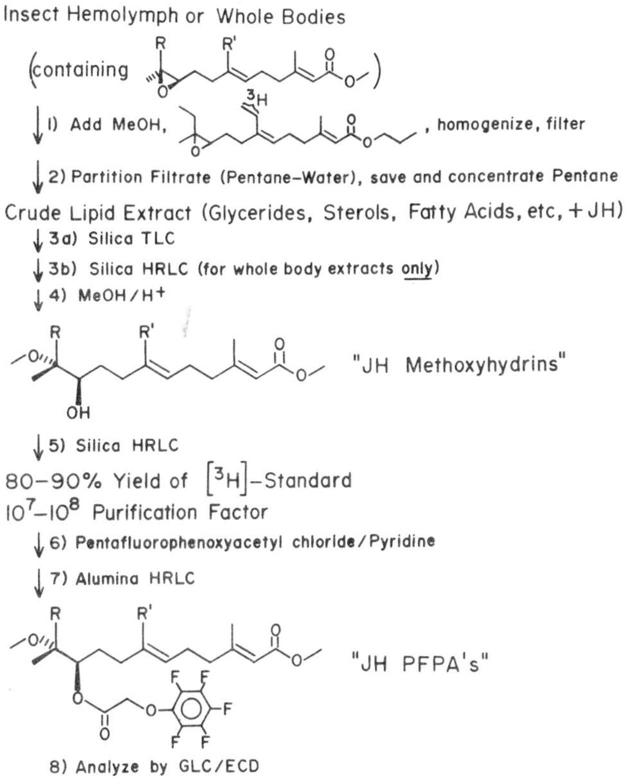

Insect Hemolymph or Whole Bodies

(containing [structure], homogenize, filter)

↓ 1) Add MeOH, [structure]

↓ 2) Partition Filtrate (Pentane-Water), save and concentrate Pentane

Crude Lipid Extract (Glycerides, Sterols, Fatty Acids, etc, + JH)

↓ 3a) Silica TLC

↓ 3b) Silica HRLC (for whole body extracts only)

↓ 4) MeOH/H+

[structure] "JH Methoxyhydrins"

↓ 5) Silica HRLC

80–90% Yield of $[^3H]$–Standard
10^7–10^8 Purification Factor

↓ 6) Pentafluorophenoxyacetyl chloride/Pyridine

↓ 7) Alumina HRLC

[structure] "JH PFPA's"

8) Analyze by GLC/ECD

Fig. 5. Scheme for processing insect samples for GLC/electron capture detection analysis. From Bergot *et al.* [55], reprinted by permission of Pergammon Press (*Life Sci.*).

extraction-partitioning steps and final GLC/ECD analysis, three chromatographic purifications and two chemical derivatizations

are required. Whole body extracts require one additional
chromatographic purification. Of many haloacids investigated,
the pentafluorophenoxyacetate moiety was found to impart excep-
tionally high electron capturing affinity, while possessing
appreciable volatility for derivatives of this molecular
weight. Detectability limits are about 300-400 femtograms with
analytical standards, while with derivatives of biological
samples this level is tenfold higher (3-4 pg). The method is
technically difficult; a skilled practitioner can analyze only
∿6-10 samples per month. A JH titer analysis of hemolymph
from diapausing *D. grandiosella* larvae (southwestern corn
borer) is shown in Figure 6. All three JHs are present, but
the predominant hormone is JH II (3 ng/ml). As in any GLC/ECD
procedure involving "tagging" of a hydroxyl group with electrophoric derivatives, specificity of analysis remains a critical
problem, especially with processed whole bodies. For optimal
results the sample should be chemically pure at the JH methoxy-
hydrin stage; however, interfering peaks arising from
hydroxylic impurities can frequently be resolved from JH derivatives by analyzing sample aliquots on several GLC stationary
phases, differing in selectivity [55].

In addition to the study of JH titers in the southwestern
corn borer [55], additional applications of our method are in
press. We find [51] that JH III predominates in two flies
(*Musca domestica* and *Sarcophaga bullata*) at low titers (<1
ng/g body weight), that JH II is by far the principal hormone
in fourth and fifth instar larvae of *Samia cynthia*, and that
JH I and JH II predominate in larval *M. sexta*, while the
balance shifts to JH II and JH III in adult females of this
species.

The GLC/ECD method of the Texas A&M group (Peter *et al.*
[70], Dahm *et al.* [34]) utilizes 2,4-dichlorobenzoate esters of
the JH methoxyhydrins and a different internal standard, a
[*methoxy*-³H]homolog of 2*E*,6*E*,10*trans* JH I bearing ethyl groups
at carbons 3, 7, and 11. Whole body analysis has apparently
not been attempted with this method. Hemolymph analyses have
been performed on several life stages of *M. sexta* [70], and
while there are certain qualitative and quantitative

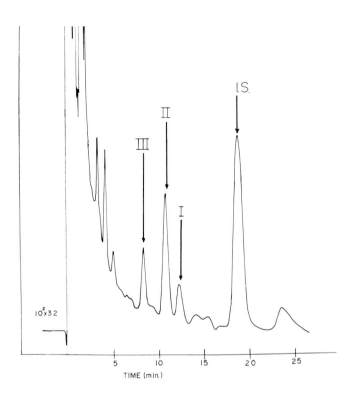

Fig. 6. Chromatogram obtained by GLC/ECD analysis of hemolymph
of diapausing larvae of *Diatraea grandiosella*. Derivatives of
the natural JHs are denoted only by Roman numerals (without
JH), while I.S. stands for the homologous internal standard
derivative. Injected masses of the four derivatives were 78,
205, 62, and 720 pg. Conditions were: 180x0.2 cm glass column,
3% OV-17 on 100-120 mesh Chromosorb W-AW-DMCS maintained at
237° isothermal, nitrogen carrier at 40 ml/min, electron capture detector (15 mCi Nickel-63, Tracor). From Bergot *et al.*
[55], reprinted by permission of Pergamon Press (*Life Sci.*).

discrepancies between their data and ours [70,51], both groups
agree on the previously mentioned shift in JH profiles from
larvae to female adult. Peter *et al.* [70] find JH II to be the
exclusive hormone in penultimate larval *H. cecropia* hemolymph.

A comparison of the methodologies of our two research
groups is of interest, despite the fact that these current
techniques may soon be vastly improved or even rendered obso-
lete. The number of procedural steps prior to acylation with
electrophoric reagents is similar. The Texas A&M method

appears slightly more laborious than that of Zoecon, since the
A&M researchers purify the final electrophoric JH derivative by
two preparative TLC steps prior to HRLC, in contrast to a
simple alumina pencil column filtration prior to HRLC in the
Zoecon procedure. A major difference lies in the HRLC purification of final derivatives: the A&M group separates individual
zones corresponding to the retention volume of each JH derivative, discarding the internal standard, and subsequently ana-
lyzes each fraction by GLC/ECD separately. The Zoecon workers
collect a band of eluent broad enough to encompass the derivatives of the internal standard and all three JHs. The former
approach is more conducive to specificity of analysis, but is
no longer an internal standard assay for final quantitation,
and precision must accordingly be sacrificed. The latter technique is more likely to give a sample containing interfering
materials if samples were not rigorously purified before derivatization, but the derivatives of all three JHs, plus standard,
will be visible in a single chromatogram.

It should be apparent from the above discussion that
there still exists a great need for improved simplicity and
specificity in GLC/ECD analysis of the juvenile hormones.
Specificity could be achieved most likely *via* exploitation of
microchemical derivatizations that are inherently more selective reactions than acylation of a hydroxyl substituent. A
motivating factor for development of improved GLC/ECD methods
is the rather modest cost of an electron capture chromatograph,
especially compared to a GLC/mass spectrometer.

Protein binding assays

The simplicity, sensitivity, and specificity of radio-
immunoassays (RIA) has contributed to their widespread adop-
tion in biochemistry, despite the relatively lower precision
of these methods compared to physico-chemical procedures.
Lauer *et al.* [80] have prepared JH antibodies by conjugation
of the *N*-hydroxysuccinimide ester of the saponification
product of racemic JH III to human serum albumin, and inocu-
lated rabbits with the resultant conjugate. Antibodies were
isolated and their specificity determined by competition experiments against JH analogs (unfortunately, eight of the fifteen

JH analog structures are missing their 3-methyl substituents as
drawn in this publication [80]). The antibodies proved to be
rather specific for JH I, II, and III, although certain analogs, especially those containing terminal epoxy groups, cross-
reacted to a certain extent. A small population of the antibodies apparently possessed somewhat higher affinity for JH I
than JH III, the immunogen. These antisera do not appear to
have been utilized in any biological investigations, perhaps
due to limited sensitivity and affinity.

In applying RIA or related techniques to JH assay, a crude
sample (such as centrifuged hemolymph) cannot be simply mixed
with the antibody-labeled hapten complex if qualitative discrimination between the three JHs is desired. To effect this
discrimination, the sample must be extracted, the bulk of lipid
removed, the residue fractionated (preferably by HRLC) into
three zones corresponding to the known JHs, and each fraction
individually assayed for JH content. Thus several of the procedural steps requisite in physico-chemical methods are also
mandated in protein binding assays, and three assays per sample
are required.

Goodman *et al.* [81] have recently developed an assay for
JH using the naturally occurring JH-specific binding protein
from *M. sexta* hemolymph. After partial purification and delip-
idation of the binding protein, [^{3}H]JH I is added. The resultant complex is mixed with a suitably fractionated hemolymph
extract containing the presumed JH to be identified. After a
16-hr incubation to assure complete equilibration, free,
unbound JH is adsorbed on dextran-coated charcoal and removed
from solution by centrifugation. Radioassay of the supernatant
determines the amount of [^{3}H]JH I remaining on the specific
binding protein, and therefore the amount of JH added in the
sample. The method is said to be rapid, but the sensitivity
is only 1 ng under optimal conditions. Hexane extracts of

hemolymph or goat serum show a "background" of 2-3 ng. The
familiar problems of adsorption of JH from aqueous solution
were encountered and were partly alleviated by Carbowax 20M
treatment of glassware. The binding protein from *M. sexta*
appears more specific than the antibodies prepared by Lauer *et
al.* [80] with respect to JH analogs, and in fact, considerable
qualitative discrimination between the three JHs is observed.
The most lipophilic hormone, JH I, has the highest affinity for
the protein, with JH II and III exhibiting 35% and 8% of the
affinity of JH I. The assay will thus have the lowest sensitivity for the poor competitor JH III, unless [^{3}H]JH III were
used as ligand. Perhaps these considerations account for the
author's statement that quantification is "extremely complex"
[81], and for their presenting only cumulative JH titers
determined by the competitive binding protein assay, expressed
(like bioassay) in terms of "JH I equivalents." Hopefully
additional development can lead to a more sensitive assay with
less problems of quantitation, since the apparatus used is
available in a well equipped biochemical laboratory.

SUMMARY

    Structural and stereochemical analysis of the three known
insect juvenile hormones is reviewed. Techniques for identifying microgram quantities of JH are discussed, the most useful appearing to be microchemical degradations or reactions
and coupled gas-liquid chromatography/mass spectrometry (GLC/
MS). The combined use of electron impact and chemical ionization MS promises to be an immensely valuable technique.
Methods for extracting and purifying JH are critically
reviewed. Purification techniques discussed include molecular
distillation, low temperature precipitation, column and thin-
layer chromatography, and partitioning systems; while special
emphasis is placed on gas-liquid chromatography and high
resolution liquid chromatography. The latter technique is
deemed the method of choice for final separation of prepuri-
fied hormone samples. Methods for determination of physiological titers of the three known JHs are reviewed according
to the mode of detection utilized--mass spectrometry (multiple

ion detection), electron capture techniques, and protein
binding assays. Current progress in the rapidly evolving field
of titer assays indicates that much improved, generally applicable methods will be available within a few years.

ACKNOWLEDGEMENTS

I am much indebted to B. J. Bergot and L. L. Dunham for
collaborating on many aspects of the analytical chemistry of
juvenile hormones. I gratefully acknowledge K. J. Judy, D. S.
King, B. J. Bergot, and G. B. Staal, whose many helpful
comments greatly improved the original manuscript. I also
thank many colleagues for allowing me to quote unpublished data
and preprints of unpublished manuscripts. I am very grateful
to C. Schubert for patiently and skillfully typing the many
drafts of this chapter. Partial financial support was provided
by the National Science Foundation Grant No. BMS 74-19048.

REFERENCES

1. C.M. Williams, Nature, 178(1956)212.

2. H. Röller, J.S. Bjerke, and W.H. McShan, J. Insect
   Physiol., 11(1965)1185.

3. H. Röller and J.S. Bjerke, Life Sci., 4(1965)1617.

4. H. Röller, J.S. Bjerke, L.M. Holthaus, D.W. Norgard, and
   W.H. McShan, J. Insect Physiol., 15(1969)379.

5. H. Röller, K.H. Dahm, C.C. Sweeley, and B.M. Trost, Angew.
   Chem. Internat. Edit., 6(1967)179.

6. K.H. Dahm, B.M. Trost, and H. Röller, J. Am. Chem. Soc.,
   89(1967)5292.

7. A.S. Meyer, H.A. Schneiderman, E. Hanzmann, and J.H. Ko,
   Proc. Nat. Acad. Sci. USA, 60(1968)853.

8. A.S. Meyer and E. Hanzmann, Biochem. Biophys. Res.
   Commun., 41(1970)891.

9. P. Loew and W.S. Johnson, J. Am. Chem. Soc., 93(1971)3765.

10. D.J. Faulkner and M.R. Petersen, J. Am. Chem. Soc., 93
    (1971)3766.

11. K. Nakanishi, D.A. Schooley, M. Koreeda, and J. Dillon,
    Chem. Commun., (1971)1235.

12. A.S. Meyer, E. Hanzmann, and R.C. Murphy, Proc. Nat.
    Acad. Sci. USA, 68(1971)2312.

13. V.B. Wigglesworth, Principles of Insect Physiology,
    Chapman and Hall, London, 7th Ed., 1972, p. 78.

14. G.B. Staal, Annu. Rev. Entomol., 20(1975)417.

15. K.H. Dahm and H. Röller, Life Sci., Part II, 9(1970)1397.

16. H. Röller and K.H. Dahm, in W.J. Burdette (Editor),
    Invertebrate Endocrinology and Hormonal Heterophylly,
    Springer-Verlag, New York, Berlin, 1974, p. 235.

17. P.D. Shirk, K.H. Dahm, and H. Röller, Z. Naturforsch.,
    1976, in press.

18. H. Röller and K.H. Dahm, Naturwissenschaften, 57(1970)454.

19. K.J. Judy, D.A. Schooley, L.L. Dunham, M.S. Hall, B.J.
    Bergot, and J.B. Siddall, Proc. Nat. Acad. Sci., USA,
    70(1973)1509.

20. K.J. Judy, D.A. Schooley, M.S. Hall, B.J. Bergot, and J.B.
    Siddall, Life Sci., 13(1973)1511.

21. B. Lanzrein, M. Hashimoto, V. Parmakovich, K. Nakanishi,
    R. Wilhelm, and M. Lüscher, Life Sci., 16(1975)1271.

22. K.H. Trautmann, P. Masner, A. Schuler, M. Suchý, and H.-K.
    Wipf, Z. Naturforsch., 29C(1974)757.

23. R.C. Jennings, K.J. Judy, D.A. Schooley, M.S. Hall, and
    J.B. Siddall, Life Sci., 16(1975)1033.

24. K.H. Trautmann, A. Schuler, M. Suchý, and H.-K. Wipf,
    Z. Naturforsch., 29C(1974)161.

25. K.J. Judy, D.A. Schooley, R.G. Troetschler, R.C. Jennings,
    B.J. Bergot, and M.S. Hall, Life Sci., 16(1975)1059.

26. M.M. Blight and M.J. Wenham, Insect Biochem., 6(1976)35.

27. P.J. Müller, P. Masner, K.H. Trautmann, M. Suchý, H.-K.
    Wipf, Life Sci., 15(1974)915.

28. G.E. Pratt and R.J. Weaver, J. Endocrinol., 64(1975)67P.

29. M.M. Blight and M.J. Wenham, J. Insect Physiol., 22(1976)
    141.

30. P. Paguia, P. Masner, K.-H. Trautmann, and A. Schuler,
    Experientia, 32(1976)122.

31. G.B. Staal, in J.J. Menn and M. Beroza (Editors), Insect
    Juvenile Hormones - Chemistry and Action, Academic Press,
    New York, 1972, p. 69.

32. J.S. Bjerke and H. Röller, in W.J. Burdette (Editor),
    Invertebrate Endocrinology and Hormonal Heterophylly,
    Springer-Verlag, New York, Berlin, 1974, p. 130.

33. J. de Wilde, G.B. Staal, C.A.D. de Kort, A. de Loof, and
    G. Baard, Kon. Ned. Akad. Wetensch., Proc. Ser. C., 71
    (1968)321.

34. K.H. Dahm, G. Bhaskaran, M.G. Peter, P.D. Shirk, K.R.
    Seshan, and H. Röller, in L.I. Gilbert (Editor), The
    Juvenile Hormones, Plenum Press, New York, 1976, in press.

35. P. Schmialek, Z. Naturforsch., 16B(1961)461.

36. P. Schmialek, Z. Naturforsch., 18B(1963)462.

37. R.D. Goodfellow and L.I. Gilbert, Amer. Zool., 3(1963)508.

38. L.I. Gilbert and H.A. Schneiderman, Science, 128(1958)844.

39. A.S. Meyer, E. Hanzmann, H.A. Schneiderman, L.I. Gilbert,
    and M. Boyette, Arch. of Biochem. and Biophys., 137(1970)
    190.

40. H. Röller and K.H. Dahm, Recent Prog. in Horm. Res., 24
    (1968)651.

41. B.M. Trost, Acc. Chem. Res., 3(1970)120.

42. L.I. Gilbert and D.S. King, in M. Rockstein (Editor), The
    Physiology of Insecta, Vol. I, Academic Press, New York,
    2nd Ed., 1973, Ch. 5, p. 250.

43. W.S. Bowers, M.J. Thompson, and E.C. Uebel, Life Sci., 4
    (1965)2323.

44. A.S. Meyer, Bull. Soc. Ent. Suisse, 44(1971)37.

45. W.S. Johnson, S.F. Campbell, A. Krishnakumaran, and A.S.
    Meyer, Proc. Nat. Acad. Sci., USA, 62(1969)1005.

46. A. Horeau and H.B. Kagan, Tetrahedron, 20(1964)2431.

47. R.J. Liedtke and C. Djerassi, J. Org. Chem., 37(1972)2111.

48. D.F. Hunt and J.F. Ryan, Anal. Chem., 44(1972)1306.

49. M. Beroza, Acc. Chem. Res., 3(1970)33.

50. E.E. van Tamelen, Acc. Chem. Res., 1(1968)114.

51. D.A. Schooley, K.J. Judy, B.J. Bergot, M. S. Hall, and
    R.C. Jennings, in L.I. Gilbert (Editor), The Juvenile Hor-
    Mones, Plenum Press, New York, 1976, in press.

52. R.J. Anderson, C.A. Henrick, J.B. Siddall, and R. Zurflüh,
     J. Am. Chem. Soc., 94(1972)5379.

53. B.D. Hammock, S.S. Gill, V. Stamoudis, and L.I. Gilbert,
     Comp. Biochem. Physiol., 53B(1976)263.

54. W.W. Christie, Lipid Analysis, Pergamon Press, New York,
     1973, p. 32.

55. B.J. Bergot, D.A. Schooley, G.M. Chippendale, C.-M. Yin,
     Life Sci., 18(1976)811.

56. W.W. Miller, J.S. Wilkins, and L.L. Dunham, J. Assoc. Off.
     Anal. Chem., 58(1975)10.

57. A.I. Vogel, A Textbook of Practical Organic Chemistry,
     John Wiley & Sons, New York, 3rd Ed, 1966, p. 120.

58. A.S. Meyer and H.A. Ax, J. Insect Physiol., 11(1965)695.

59. M.A. Bieber, PhD Dissertation, Michigan State University,
     1973, p. 63.

60. Pharmacia Fine Chemicals, Technical Bulletin on Sephadex
     LH-20 (1970).

61. K.J. Kramer, L.L. Sanburg, F.J. Kezdy, and J.H. Law,
     Proc. Nat. Acad. Sci., USA, 71(1974)493.

62. G.E. Pratt and S.S. Tobe, Life Sci., 14(1974)575.

63. J. Nowock, W. Goodman, W.E. Bollenbacher, and L.I.
     Gilbert, Gen. Comp. Endocrinol., 27(1975)230.

64. L.L. Sanburg, K.J. Kramer, F.J. Kezdy, and J.H. Law,
     J. Insect Physiol., 21(1975)873.

65. L.I. Gilbert and H. I. Schneiderman, Am. Zool., 1(1961)11.

66. A.S. Meyer and H.A. Ax, Anal. Biochem. 11(1965)290.

67. D.A. Schooley, K.J. Judy, B.J. Bergot, M.S. Hall, J.B.
     Siddall, Proc. Nat. Acad. Sci., 70(1973)2921.

68. R.C. Jennings, K.J. Judy, and D.A. Schooley, Chem.
     Commun., (1975)21.

69. R.D. Goodfellow, Y.-S. Huang, and J.-C. Wu, Amer. Zool.,
     14(1974)1291.

70. M.G. Peter, K.H. Dahm, and H. Röller, Z. Naturforsch.,
     1976, in press.

71. L.L. Dunham, D.A. Schooley, and J.B. Siddall,
     J. Chromatogr. Sci., 13(1975)334.

72. K. Grob, Chromatographia, 5(1972)3.

73. A.S. Meyer and J.Z. Knapp, Anal. Biochem., 33(1970)429.

74. F.A. Fitzpatrick and S. Siggia, Anal. Chem., 45(1973)2310.

75. K. Nakanishi, Proc. XXIII Int. Congr. of Pure and Appl. Chem., 3(1971)27.

76. L.R. Snyder and J.J. Kirkland, Introduction to Modern Liquid Chromatography, John Wiley & Sons, New York, 1974, Ch. 8, p. 239.

77. J.J. Kirkland, J. Chromatogr., 83(1973)149.

78. M.A. Bieber, C.C. Sweeley, D.J. Faulkner, and M.R. Petersen, Anal. Biochem., 47(1972)264.

79. L.W. van Broekhoven, A.C. van der Kerk-van Hoof, and C.A. Salemink, Z. Naturforsch., 30C(1975)726.

80. R.C. Lauer, P.H. Solomon, K. Nakanishi, and B.F. Erlanger, Experientia, 30(1974)558.

81. W. Goodman, W.E. Bollenbacher, H.L. Zvenko, and L.I. Gilbert, in L.I. Gilbert (Editor), The Juvenile Hormones, Plenum Press, New York, 1976, in press.

CHAPTER 8

ANALYTICAL BIOCHEMISTRY OF INSECT NEUROTRANSMITTERS AND THEIR ENZYMES

THOMAS SMYTH, JR.

Department of Entomology, Department of Biochemistry and Biophysics
The Pennsylvania State University

Address: 2 Patterson Building, University Park, PA 16802   USA

CONTENTS

I. INTRODUCTION

1. Neurotransmitters

A chemical transmitter is a substance that is produced in a nerve cell and that can be released at synaptic junctions to excite or inhibit post-junctional cells. Although this definition is straightforward, the un-equivocal demonstration of the identity of a transmitter is never easy; numerous criteria must be met before a substance can be accepted as a transmitter [1]. An action potential in a given prejunctional cell may allow the release of 10^4 to 10^6 molecules of transmitter into a very confined space, usually protected by a cellular sheath. Effective mechan-isms of active uptake or enzymatic hydrolysis then effectively sequester or destroy perisynaptic transmitter within milliseconds or tens of milli-seconds. Even when one can study the total transmitter released by parallel similar neurons, the small size of insects limits the numbers of such neurons and hence the size of the transmitter pool. Nevertheless, analytical methods have been developed which can overcome these formidable problems.

The presence of a known transmitter substance in nervous tissue does not prove it to be serving as a transmitter there. Amino acid transmitters such as glutamate, aspartate and glycine have many other important bio-chemical roles; the aromatic amines serve also as precursors for cuticular tanning reagents and pigments; several putative transmitters are biochemically related, being precursors to or products of other trans-mitters. Dopamine, norepinephrine, epinephrine and octopamine form one such family. Gamma-aminobutyric acid is derived from glutamic acid and its enzymatic destruction can regenerate glutamate by transamination.

The conventional concept of the action of a transmitter is that it combines with a postjunctional membrane bound receptor molecule that either is or is coupled   to the gate of a ion selective channel through the membrane. Thus, transmitter action increases membrane permeability and membrane current and lowers membrane resistance. Which ion or ions are active (allowed to pass through the channel) determines how the membrane potential will be affected and whether the result will be excitation or inhibition.

Recently, other sorts of membrane responses have been described. Excitation or inhibition may be associated with increased membrane resis-tance rather than decreased, may last over extended periods of time measured in seconds, minutes or hours rather than milliseconds, may depend on additional classes of membrane bound receptors, and may be mediated through

second messengers such as cyclic AMP and GMP. These responses are sometimes evoked by conventional transmitters but are often caused by a variety
of other substances including specific peptides, prostaglandins, and
certain hormones. Some of these substances come from immediately
presynaptic cells, others from distant release sites. In the latter case
the substances are clearly not transmitters but may be classed as modu-
lators. Between transmitter and modulator there is a hazy area where
definations will be arbitrary. Although most of the work to date on these
newer complications has been done on molluscs and vertebrates, there are
early indications of similar synaptic complexity in insects.

In this chapter discussion will be limited to the substances for which
there is considerable evidence of conventional transmitter function [1-5].
Several other possible transmitters are chemically related to members of
this primary group. Similar methods of analysis will generally be appropriate for them.

2. General comments on methods for analysis

Several recent books describe analytical methods for neurotransmitters
and their enzymes [6-8]. New techniques appear most commonly in the
periodicals Journal of Neurochemistry and Analytical Biochemistry. A good
general introduction to neurotransmitters and the analytical procedures
used for them is the little book by Cooper, et al. [9].

Complete investigation of a transmitter includes also the study of the
enzymes that make it, enzymes or specific transport systems that destroy or
remove it, post-junctional receptors, precursors and metabolites. Questions
asked by the neurobiologist often may be answered by investigation of one
of these other components rather than analysis for the transmitter itself.
Cells making a transmitter should contain the appropriate enzyme, those
acted on by it should have receptors, and cells in the vicinity of the
synapse should be able to destroy or sequester it. Knowledge of these
components also is essential when analytical procedures are employed that
involve biochemical conversions as in making radioactive transmitters or
converting transmitters to products that are more easily separated or
assayed.

Until recently the only highly sensitive analytical procedures for
transmitters were bioassays which can resolve ng or pg (10^{-9} - 10^{-12} gram)
quantities. These are inexpensive in terms of equipment but are nonspecific, slow and often capricious. Experience and skill at this older
style of pharmacology are necessary if consistent results are to be

obtained. Bioassays are still used in situations where modern analytical chemical facilities are not available. At the other end of the analytical spectrum, modern mass fragmentography coupled with gas chromatography can resolve complex mixtures and may detect quantities in the fmol (10^{-15} mol) range. A proficient chemist and some very expensive equipment are required. Intermediate in cost, sensitivity and specificity are a large number of other techniques. New procedures have been proposed not only to improve sensitivity and specificity but particularly to reduce the time and labor involved in an assay and to make it possible to run large numbers of samples in a short time. One or two step extractions or separations and radio-chemical or fluorometric measurements are popular. Often these are combined with an enzymatic conversion which provides added specificity. Another recent achievement has been the development of procedures for assay of many of the commonly accepted transmitters and related compounds on single micro-samples [10]. The choice of analytical procedures will depend on the equipment and personnel available and on the sorts of biological questions to be answered.

Because transmitters have rapid biochemical turnover, especially when released from storage, it is usually necessary to stop enzyme action very rapidly. This can be done by rapid freezing of small insects or parts of them or by rapid heating by boiling or with microwave radiation. Enzyme inactivation is followed by procedures for extraction, concentration, separation, detection and measurement. It is occasionally possible to omit some of these steps. For example, certain perfusates and extracts can be bioassayed directly.

The procedures mentioned in the sections that follow either have been used for insect material or are recently developed methods that will probably prove useful in the future.

II. ACETYLCHOLINE AND ITS ENZYMES

Acetylcholine (ACh) is synthesized from choline and acetyl-coenzyme A by the enzyme choline acetyltransferase (ChAT) which is also widely known as choline acetylase (ChAc). It can be hydrolyzed by specific acetylcho-linesterases (AChE), by other cholinesterases (ChE) which have higher affinity for butyryl- or propionyl- than for acetylcholine, and by other less specific esterases to yield choline and acetic acid.

1. Acetylcholine [11]

ACh in tissues is mostly in bound or stored form. On release it is highly susceptible to hydrolysis by enzymes or by strongly acidic or basic

conditions. Therefore, the extraction procedure must rapidly inactivate
enzymes and keep them inactivated. ACh is most stable at pH 4-5.

Extracts are usually prepared from insects or tissues that have been
killed by boiling 1/2 to 2 min at about pH 4 in 0.02 N HCl [12] or in 2%
trichloracetic acid which is added after boiling [13]. The tissues are next
iced and homogenized and then centrifuged to throw down precipitated
proteins. This may be done either in the boiling medium or in 1 N formic
acid-acetone (15:85) by volume. Residues may be resuspended in extraction
medium and recentrifuged twice to assure complete recovery of ACh. Quick
frozen samples may be pulverized and then extracted [10]. HCl extracts may
be neutralized and prepared in a physiological saline for direct bioassay.
However, a separation procedure is usually employed next. ACh can be
extracted into an organic phase as an ion pair with a hydrophobic anion such
as tetraethyl boride [14]. Concentration from aqueous solution can be by
simple evaporation (with considerable risk of loss of activity) or by
precipitation of sparingly soluable salts (reineckate or periodate) with an
added co-precipitant such as choline to insure quantitative recovery [11].
The formic acid-acetone extraction fluid can be removed by evaporation with
little loss of activity.

Separation of ACh from concentrated extracts can be accomplished in
several ways, but the method should be capable of separating it from
choline-containing membrane lipids. Paper chromatography on Whatman No. 1
or 4 filter paper, ascending or descending, can be used. A solvent system
that works well is ethyl acetate-pyridine-water (50:30:20) by volume [13].
Good separation can also be accomplished on a Sephadex LH-20 molecular
sieve column [12]. The Sephadex is suspended in 10^{-3} N HCl overnight and
then sedimented into a 2 x 20 cm column. Extracts are added above a filter
paper disc on top of the gel. The column is run with a head of 10^{-4} N HCl
at 0.25-0.3 ml/min.

Paper electrophoresis can be used to separate a variety of compounds
related to ACh [12,15]. Samples are applied to the centers of Whatman No. 1
paper strips and air dried. Electrophoresis is carried out at room temperature at 18 V/cm for 1 hr at pH 2. The buffer, 1.5 M acetic acid-0.75 M
formic acid, is applied to both sides of the streak of sample. This buffer
is completely volatile and also can be used to release bound ACh from tissue
fragments applied to the paper. Another electrophoretic system has recently
been proposed [16].

When relatively large amounts of choline esters have been separated they can be detected with the following reagents [17]: Spray reagent. 0.2 ml 0.5 N Mg dipicrylamine, 50 ml methanol, 49 ml distilled water, 1 ml concentrated NH_3. ACh and choline:blue-red. Dipping reagent. 2% aqueous phosphomolybdic acid, 1 min. Wash in butanol, 5 min. Wash in water, 5 min. Dip into 0.4% stannous chloride in 3 N HCl. Blue spots on white background.

Small amounts of ACh are easier to locate if the transmitter has been isotope labeled. Alternatively, addition of small amounts of [^{14}C] or [^{3}H] ACh to the cold sample can indicate the location of the ACh and also provide information about losses during the isolation procedure. Scintillation counting is both sensitive and quantitative. Another type of radiochemical assay is based on the enzymatic hydrolysis of [^{3}H] acetate labeled ACh by AChE, the acetate produced being counted. When a sample containing ACh is added, less labeled acetate is produced per unit time [18].

Bioassays are much more sensitive than the older chemical methods and are still in use in laboratories where neurobiologists are not equipped to perform the newer chemical analyses. Since the bioassays are nonspecific, several sorts of controls must be run. It is customary to destroy the ACh in part of the sample by raising the pH to 11 with 0.3 N NaOH which is then neutralized before assay. Another way to destroy the ACh is by incubation with AChE which is inactivated before assay. Selection of an appropriate cholinergic blocking agent should prevent the response of the assay preparation. Finally, it is usual to employ more than one kind of assay or analysis. Since bioassays are quite sensitive to many tissue components and drugs, and especially to potassium ions, preliminary cleanup is often necessary. Some popular bioassay systems are frog rectus abdominis muscle [19,20], guinea pig ileum [11], clam heart [21], and leech muscle [19,22]. Frogs are more sensitive in winter than in summer; summer frogs should be held at 2-3°C for a few days before testing. Clams should be freshly collected. Several species can be used. Although the leech muscle assay was worked out with the European medicinal leech, other species also can be used. Sensitivity is improved if the muscle is small and is suspended in a very small bath or is superfused drop-wise. Recovery time is shorter if only isometric responses are permitted. Concentrations are estimated by interpolation from semilog plots of responses to ACh standards giving higher and lower responses than the unknown.

Fellman [23] has reported a sensitive fluorometric procedure for assay for ACh in which hydrazine is reacted with the ester group and the acetylhydrazide produced is then coupled with salicylaldehyde to yield the intensely fluorescing hydrazone. This seems not to have been used for insect material.

For small samples gas chromatography affords the best method of separation. Although the salts of ACh are not volatile, it is possible to make volatile derivatives which can then be chromatographed. The ester group can be separated as acetic acid or ethanol, or a methyl group can be removed from the quaternary nitrogen to produce the volatile tertiary amine [11]. Volatile products can also be produced by pyrolysis. The most sensitive procedures now available combine pyrolysis gas chromatography with mass spectrometry [24,25].

2. Choline acetyltransferase (Choline acetylase)

ChAc should be present in all cholinergic neurons and its distribution correlates well with that of ACh in the insect nervous system [26]. It is now usually determined by a radiochemical micromethod [27,28] in which tissue homogenate is incubated with choline chloride or bromide and [^{14}C] or [^{3}H] acetyl-CoA to form labeled ACh. An anti-ChE such as eserine, (10^{-5}M), is included to minimize destruction of this product. The ACh is precipitated as the reineckate which is washed and then dissolved in acetone or extracted with ketonic sodium tetraphenylboron for scintillation counting. This method is sufficiently sensitive that it has been successfully applied to the somata of single insect motorneurons [29].

Histochemical localization of ChAc has been attempted by reacting the -SH group of the CoA released during ACh synthesis with lead nitrate to form a mercaptide which precipitates. When this is treated with sodium sulfide, lead sulfide is formed and this is visible under a microscope [30]. The specificity of this reaction has been challenged [31] and so has the specificity of an immunohistochemical localization procedure [32].

Vertebrate ChAc has been electrophoretically separated and can be detected with a stain that involves the reduction of a tetrazolium salt (MTT) by CoA at alkaline pH [33].

3. Acetylcholinesterase and other cholinesterases

The cholinesterases of insects have been very extensively studied, largely because of the great importance of the organophosphorus and carba-mate antiesterases as insecticides. The enzymes have been purified and the kinetics of inhibition by numerous antiesterases have been studied. Several histochemical and electron microscopic techniques have been used to demonstrate the location of AChE activity. Useful recent reviews consider

the biology of cholinesterases [34], general analytical methods [35] and
changes in cholinesterases during insect development [36].

Tissue homogenates have been prepared for AChE assay in many ways.
The following procedure is but one recent example that allows the easy
preparation of soluable AChE from fly heads [37]. Flies are frozen and
shaken vigorously to detach the heads [38]. These can be separated from
bodies and smaller fragments by graded seives. They are then homogenized
in a glass homogenizer in 0.1 M sodium phosphate buffer, pH 7.55. The
concentration is adjusted to 50 heads/ml and the crude brei is spun down in
a refrigerated centrifuge at 25,000 x g for 1 hr. The supernatant can be
used without further treatment and can be kept frozen for some months without loss of activity. Some other preparative methods that have recently
been used for insect cholinesterases are given in the following references
[39-41]. The various esterases can be separated by starch [42] or polyacrylamide gel [39,40] electrophoresis.

ChE activity has been detected by a diversity of tests, some with
little modification over the last twenty years. One group of assays depends
on the acetic acid released. If ACh is hydrolyzed in a bicarbonate buffer
system, CO_2 is evolved in amounts equivalent to the acetate produced. This
can be estimated manometrically with an accuracy of 2-3% [35,43]. The acid
produced changes the pH of the reaction mixture. This can be followed
electrometrically with a pH meter, or with a pH indicator dye such as
bromthymol blue or phenol red. Determination of the amount of acid produced
by titration with standard alkali has frequently been used in studying the
activity of insect AChE [44,45]. Although manual titration using an indicator dye or potentiometer was common in the past, the task can now be
easier and more accurate if an automatic recording titrator is used [35].

Another group of methods depends on the amount of choline released --
or thiocholine if acetyl- or butyrylthiocholine is used as the substrate.
The most popular procedure at present is that of Ellman, et al. [46] in
which acetylthiocholine is hydrolyzed by AChE. The thiocholine produced is
reacted with the sulfhydryl reagent dithiobisnitrobenzoic acid (DTNB) to
form the yellow amine of 5-thio-2-nitro-benzoic acid. This is measured
photometrically at 412 nm.

The amount of ACh remaining in the reaction mixture also can be determined. In vitro assays of ChE activity generally employ much larger amounts
of ACh than is normally present in tissue extracts. Therefore, less
sensitive analytical methods can be used. A simple colorimetric assay
introduced by Hestrin [47] reacts the remaining ester with alkaline

hydroxylamine to produce a hydroxamic acid. This forms a red-purple complex with $FeCl_2$ in acid solution and can be measured spectrophotometrically at 540 nm.

ChE activity can also be followed radiometrically using acetate labeled substrate and observing either the ACh remaining [48] or the labeled acetate produced [49,50].

No ChE is absolutely specific for choline esters. Therefore it is possible to provide substrates that yield products for which there are simple or sensitive assay techniques. Phenyl, nitrophenyl, naphthyl, indoxyl and indophenyl esters have been used [35].

The demonstration of a high local concentration of AChE activity in tissue is often taken as an indicator of cholinergic function, although the ubiquitous distribution of enzymes with ChE activity and the limited specificity of the histochemical procedures makes such a demonstration much less than a sure proof. The Koelle-Friedenwald technique or one of its many modifications is generally used for the histochemical demonstration of ChE [51-55]. Frozen sections or whole organs are incubated with acetylthiocholine in the presence of copper glycinate. The thiocholine released by hydrolysis precipitates as copper thiocholine. To prevent this from dissolving, incubation is carried out in the presence of a small amount of added copper thiocholine. Treatment with ammonium sulfide converts the precipitate to copper sulfide. Although this procedure has been carried out on formaldehyde fixed material, enzyme activity is reduced. To distinguish between the various esterases, inhibitors can be introduced into the incubation medium. Eserine (physostigmine) and related carbamates inhibit both AChE and the other ChE. Organophosphorus inhibitors are more effective against the non-specific ChE than against AChE. For vertebrate material DFP is reported to be the best selective inhibitor for histochemical use [56], but it is hazardous to use. Paraoxon, iso-OMPA and other organophosphorus compounds have also been used. These are safer to use but do not discriminate as well between the classes of cholinesterases. However, in insect central nervous systems it appears that AChE activity is high relative to ChE, so the problem is to discriminate between AChE and other classes of esterases [55]. Carboxylesterases (aliesterases) are also inhibited by the phosphates but not by carbamates; arylesterase is resistant to these inhibitors [57]. BW 284 C51 (Burroughs Wellcome & Co.) is reported to be a specific inhibitor for AChE [58] and has been used with insect esterases [59]. Histochemical discrimination among the esterases can also be facilitated by the use of alternative substrates. AChE more effectively hydrolyzes acetylthiocholine; other ChE may prefer propronylthiocholine or butyrylthiocholine [45].

Certain less specific histochemical methods also have proved useful. Wigglesworth [60] preferred an indoxyl method.

For localization of AChE activity in electronmicrographs the thiocholine technique has been modified by Karnovsky [61]. The thiocholine produced reduces ferricyanide to ferrocyanide which then reacts with copper to form fine copper ferrocyanide deposits. An alternative method proposed by Barnett [62] uses thiolacetic acid as substrate and lead nitrate as the capturing reagent. The tissue can be incubated directly or prefixed with cold 2.5% glutaraldehyde at pH 7.2 in 0.05 M cacodylate buffer. Smith and Treherne [63] fixed cockroach ganglia 20-60 min and then washed in cold cacodylate buffered 0.25 M sucrose for up to 24 hr before incubation. Incubation was terminated by postfixation with osmium tetroxide.

III. AMINO ACID TRANSMITTERS AND SELECTED ENZYMES

Analytical methods for amino acids are discussed in detail in another chapter in this book. Methods for isolation and determination of amino acid transmitters in vertebrate brain have been reviewed by Gaitonde [64]. The amino acids of special interest as neurotransmitters in insects are L-glutamate (Glu) which almost certainly mediates excitatory somatic neuromuscular transmission and γ-aminobutyrate (GABA) which is an inhibitory transmitter [1-3]. Glu and also L-aspartate have been suggested to be transmitters at insect visceral nerve-muscle synapses [65]. Glu serves many biochemical functions, can be formed by several pathways and is relatively abundant in all cells and in hemolymph. However, the extracellular concentration in the vicinity of glutamergic synapses is apparently kept low by an effective and specific uptake mechanism [66,67].

Decarboxylation of Glu by glutamic acid decarboxylase (GAD) produces GABA. GAD (Type I) and GABA are essentially limited to cells using GABA as a transmitter. By contrast, the major enzyme system capable of destroying GABA is widely distributed. The mitochondrial enzyme GABA-transaminase (GABA-T) can convert GABA and α-oxoglutarate to succinic semialdehyde and Glu. Succinic semialdehyde dehydrogenase (SSADH) then oxidizes the semialdehyde to succinate which enters the Krebs Cycle. GABA can also be converted to a variety of other products. However, uptake mechanisms for GABA [68,69] are effective at GABAergic synapses and probably predominate over metabolic conversions in removing perisynaptic transmitter.

1. L-Glutamate and gamma-amino butyrate

Tissues and perfusates to be analyzed for Glu, GABA and related compounds are generally quickly frozen and then lyophyllized, homogenized in
an ice bath or pulverized while frozen. Deproteination and extraction of
amino acids may be accomplished by addition of 5-10% trichloracetic acid,
0.4 M perchloric acid, 1 M acetic acid followed by acetone, or 75% ethanol,
the latter being recommended in preparation for enzyme microassays. Tissue
samples are centrifuged and the pellets are washed to recover all the amino
acids.

The individual amino acids are commonly separated by paper chromatography [70], thin layer chromatography [71] or by selective absorption on
columns [64]. Amberlite CG 50 resin retains GABA; AG1-X8 resin retains
glutamate and aspartate. Paper electrophoresis is also reported to effect
good separation [72]. Gas-liquid chromatography can be used to separate
and quantitate DPN-amino acid methyl ester derivatives and has the advantage of very high sensitivity [73, 74]. Automatic amino acid analyzers
afford a particularly convenient means of analysis [75, 76].

Location of amino acids on paper after chromatography or electrophoresis and quantitation of large samples can be accomplished visually and
spectrometrically with ninhydrin [64]. Much greater sensitivity can be
achieved with dansyl derivatives of the amino acids, especially when combined with radiometric techniques [77].

The most commonly used microassays employ the enzymatic-fluorometric
technique of Graham and Aprison [78]. Glu in the tissue extract is oxidized
to α-ketoglutarate by glutamic acid dehydrogenase with concommitant quantitative conversion of NAD^+ to NADH which is measured by its native fluorescence. Similarly, GABA can be enzymatically converted to succinate by
GABA-T and SSADH with formation of the highly fluorescent NADPH [79]. A
recent report suggests a rapid radioreceptor assay for GABA [80]. Synaptosomes are prepared with tritiated GABA. When the sample is added, any
GABA it contains competes with the labeled GABA for binding sites, releasing a proportional amount of radioactivity.

The high affinity uptake systems can also serve as indicators of
specific types of synapses. By scintillation counting Fader and Salpeter
[66] found a greater accumulation of tritiated Glu following stimulation
of a nerve-muscle preparation. Autoradiography showed this to be concentrated at the neuromuscular junctions, particularly in sheath cells.

2. Enzymes

Enzymes concerned with Glu metabolism in fleshfly flight muscle have
been extensively surveyed by Donnellan, et al. [81]. Their efforts to
isolate synaptosomes and associate particular Glu enzymes with these were
not successful.

GAD assays have used uniformly ^{14}C labeled glutamate and followed the
production of labeled CO_2 [82, 83] or labeled GABA [82, 84]. As in vertebrates, there appear to be two kinds of GAD, a general mitochondrial enzyme
and another characteristic of synaptosomes or nervous tissue [83]. The
location of the latter in the nervous system is more likely to reflect sites
of GABA synthesis than glutamergic synapses.

The GABA-T and SSADH system is present in insects [85] but has not been
extensively studied.

IV. AROMATIC PRIMARY AMINES AND THEIR ENZYMES

The phenyl-, catachol- and indolamines and histamine are grouped
together because of similarity in their biosynthetic and degradative enzymes
and because they are frequently extracted together and assayed by the same
or similar methods. They are often called the "biogenic amines" although
this term is also used in a much wider sense.

The aromatic amino acids phenylalanine and its 4-OH derivative tyrosine
can serve as precursors for several candidate transmitter substances.
Tyrosine can be decarboxylated to form tyramine which is then β-hydroxylated
to yield octopamine. This is present in the firefly light organ and may
serve there as a transmitter [86]. Ring hydroxylation at the 3 position
converts octopamine to the catecholamine norepinephrine (NE), also known
as noradrenalin. More commonly, NE is derived from tyrosine by a different
sequence of enzymatic reactions: ring oxidation to dihydroxyphenylalanine,
decarboxylation to dopamine (DA) and then β-hydroxylation to NE. NE, in
turn, can be N-methylated to form epinephrine (=adrenalin). DA is more
abundant than NE in insects, and if epinephrine is present, it is in very
small amounts [87-89]. Most or all of these substances can serve as precursors for tanning reagents [90]. However, DA and NE tend to be particularly abundant in parts of the nervous system [91], especially in certain
neurons that have catecholamine type dense core vesicles [92], and they are
reported to have actions on heart, visceral muscle and salivary glands that
suggest transmitter function [2, 93, 94].

Analogously, the amino acid tryptophan, an important pigment precursor, can undergo ring hydroxylation at the 5 position followed by

decarboxylation to produce 5-hydroxytryptamine (5-HT) which is also known
as serotonin. 5-HT has been found in insect nervous systems [93, 95] but
the evidence for transmitter function is weak.

Again analogously, histidine can be decarboxylated to form histamine.
This is a component of honeybee venom [96] but there are no indications that
it serves as a transmitter in insects.

Termination of transmitter action by these putative transmitters in
vertebrates is believed to be by a rather nonselective re-uptake mechanism
[97] rather than by enzymatic destruction. The same is likely to hold true
for insects. Selective uptake of indolamine in desert locust brain has
been reported by Klemm and Schneider [95]. However, there are generally
distributed nonspecific enzymes capable of inactivating the aromatic primary amines. Monoamine oxidase (MAO) hydroxylates the beta carbon.
Catecholamines may also be methylated at the 3 position by catechol-ortho-
methyltransferase (COMT). MAO and COMT may work sequentially in either
order.

1. Catecholamines [98, 99]

Until recently bioassays were the only analytical methods with sufficient sensitivity for analysis of small samples. As with bioassays for
ACh, they are rather slow, require experience, and are nonspecific. Therefore, a variety of assays and controls are required. Östlund used two
bioassay systems to determine NE and epinephrine in several kinds of
insects [87]. NE is about twice as effective as epinephrine in elevating
cat's blood pressure but only about 1/30 as effective in relaxing fowl's
rectal caecum. Subsequent analyses of insects have employed chemical
methods.

Tissue samples are generally rapidly frozen and then homogenized in
cold 0.4 N perchloric acid or 10% trichloracetic acid. Low temperature
centrifugation or filtration removes precipitated proteins. Catechol
compounds in the supernatant or filtrate, or in a collected fluid, can
then be absorbed on alumina at slightly alkaline pH. After washing the
alumina column with water, the catecholamines can be eluted with 0.2 N
acetic acid [98]. The catecholamines can also be absorbed from tissue
extracts or fluids on ion exchange resins. If Dowex 50 columns are used,
1 N HCl elutes epinephrine and NE; 2 N HCl elutes DA. Amberlite CG 50
buffered at pH 6.5 absorbs both catecholamines and their 0-methylated
derivatives; catecholamines are eluted with 4% boric acid, methylated
amines with 2 N H_2SO_4 [98].

Separation of catecholamines can also be accomplished by other means. Paper chromatograms [87] can be developed with n-butanol saturated with 1 N HCl or with phenol saturated with 0.1 N HCl (under N_2). The catecholamines are UV flourescent; 1-2 μg is visible. Ninhydrin shows dihydroxyphenylalanine as a blue spot. DA and NE more slowly appear pale rose. If the ninhydrin treated paper is then sprayed with 0.44% potassium ferricyanide at pH 7.7 and dried at 60-70°C, DA is brownish blue, NE purple and epinephrine rose.

Dansyl derivatives are more intensely fluorescent, allowing detection of smaller samples under UV light. Dansylation combined with microchromatography on 3 x 3 cm polyamide sheets and separation with two or three successive solvent systems can resolve picomole quantities [100]. Useful solvent systems are first water/formic acid (10:3) by volume, then benzene/acetic acid (9:1) by volume, finally ethylacetate/methanol/acetic acid (2:1:1) by volume. If radiolabeled dansyl derivatives are prepared, they can be eluted for scintillation counting.

High pressure cation exchange chromatography has recently been recommended for biogenic amines [101]. Sensitivity is reported to be 1-5 nmol.

Useful procedures for gas chromatographic separation and quantitation of catecholamines have only recently been worked out. Trifluoroacetyl and pentafluoropropionyl derivatives have been tried. Possibly the most promising method is that reported by Abramson, et al. [102] using trichloromethylsilane derivatives and functional group mass spectrometry to achieve 10-100 femtomole sensitivity. This method is applicable for both primary amines and α-amino acids, and therefore can be used for all the aromatic primary amine and amino acid transmitters.

Catecholamines in fluids or tissue extracts are most frequently determined by the trihydroxyindole method in one of its many variations. The method has been reviewed by von Euler [103] and Häggendal [104]. Detailed recommended procedures can be found in chapters on methods for catecholamines [98,99]. In this method this catecholamine is first oxidized (by iodine, mercaptoethanol or ferricyanide with Cu^{++}) and then rearranged in alkaline solution to form an indole compound which is intensely fluorescent. The conditions for maximal fluorescence are somewhat different for the various products and optimal procedures have been worked out for NE and epinephrine [105] and DA [106]. The method is also useful for other catecholamines and for phenylamines such as tyramine and octopamine [99].

Another fluorescence method involves the oxidation of catechol compounds to the corresponding quinones which are then condensed with

ethylenediamine to form fluorescent products. Since a variety of catechol
compounds yield products with similar fluorescence, preliminary separation
is quite important. A detailed procedure is given by Weil-Malherbe [98].

Sensitive radiochemical assay procedures are also available. The
tissue can be provided with radioactive precursor and allowed to make
labeled transmitter [107] which is then extracted, separated and counted.
Radioenzymic conversion of extracted material is also possible [108].
Usually, COMT is used to transfer a radiolabeled methyl group from
S-adenosyl-L-methionine-methyl-^{14}C to the oxygen on the number 3 carbon of
the catecholamine. The methylated derivatives are then separated and
counted [109-112].

Most studies of insect catecholamines have used the Falck-Hillarp
histofluorescence technique. Procedural details are given by Falck and
Owman [113] and Björklund, et al. [114]. The tissue to be examined is
dissected out and quickly frozen in liquid propane at the temperature of
liquid nitrogen [115]. It is then freeze-dried and exposed to formaldehyde
gas obtained from paraformaldehyde previously kept at 70% relative humidity
[116]. The tissue can then be paraffin embedded and sectioned for fluorescence microscopy or microspectrofluorimetry. Catecholamines, phenylethyl-
amines, indoleamines and histamine produce condensation products with
different fluorescence characteristics and thus can be distinguished.
Within these chemical groups identification is more tenuous, being based on
differential rates of fluorescence development or fading.

An alternative fluorescence histochemical procedure in which the tissue
is reacted with glyoxilic acid is reported to be more sensitive [117].

2. Phenylamines [118]

Compared with the catecholamines, the phenylamines or phenolicamines
have not been very extensively studied. Tyramine and octopamine can be
extracted and separated by most of the methods used for catecholamines, a
major exception being absorption by alumina which is selective for catechol
compounds. Deproteination and extraction are usually effected in acid
solution (for example, in 0.4 N perchloric acid). This is followed by
separation by paper, thin layer or column chromatography or paper electrophoresis, as with catecholamines.

The older detection methods have low sensitivity. Recent analyses for
octopamine have mostly used the enzyme radiochemical assay of Molinoff and
Axelrod [119] in which a [^{14}C] methyl group is transferred from labeled
S-adenosyl methionine to the nitrogen of octopamine. The synephrine formed
is extracted into a mixture of isoamyl alcohol and toluene and counted

[120,121].

Another sensitive procedure is to make a preliminary separation by
paper chromatography, prepare dansyl derivatives, further separate these by
thin layer chromatography, elute and then measure fluorescence [118].

3. 5-Hydroxytryptamine [99,122]

The biologically active substances known as serotonin and enteramine
were eventually discovered to be 5-HT. Sensitive bioassays exist, notably
rat fundus strip and various invertebrate smooth muscles [123]. Recent
studies including those on insects have used chemical or histochemical
assays.

Deproteination and extraction of powdered frozen tissue can be accomplished with cold 1 N formic acid/acetone (15:85) by volume, followed by
centrifugation with resuspension in fresh extraction medium and recentri-
fugation to extract all the 5-HT. This is followed by a wash with heptane/
chloroform (8:1) by volume which is discarded. The samples are dried under
nitrogen and stored at low temperature [10]. Alternatively, tissues can be
homogenized in iced 0.4 N perchloric acid, centrifuged, and the supernatent
adjusted to pH 10.0 with NaOH. After addition of pH 10.0 borate buffer and
NaCl to saturation the 5-HT can be extracted into butanol. This is the
usual procedure in preparation for detection by the fluorescence procedures
[99].

Separation can be accomplished with paper or thin layer chromatography.
For paper, n-butanol/acetic acid/water (12:3:5) by volume, and n-propanol/
ammonia/water (100:5:10) by volume are widely used solvent systems and the
chromatograms can be run ascending or descending. For Silica Gel G coated
thin layer plates methyl acetate/isopropanol/25% ammonia (45:35:20) is a
good solvent. Desalting prior to separation is important [122].

Gas-liquid chromatography has also been used for separation of indoleamines. To form indoles with appropriate vapor pressure they can be reacted
with pentafluoropropionic anhydride [124]. Separation can be followed by
mass spectrometric assay.

The most widely used method for detection of 5-HT is that proposed by
Bogdansky, et al. [124] which observes the native fluorescence of 5-HT.
This is prepared by extraction into butanol from salt saturated alkaline
tissue homogenate followed by return to an acid aqueous phase.

A tenfold increase in sensitivity can be achieved by reacting the
extracted 5-HT with ninhydrin [125]. Following butanol extraction the 5-HT
is transferred to phosphate buffer at pH 7.0, n-heptane is added, and 5-HT

returns to the aqueous phase. Ninhydrin is then added, the mixture warmed, and after an hour fluorescence is measured. Full details appear in recent reviews of analytical methods [99,122].

Enzyme radiochemical assays depend on a two step conversion to isotope-labeled melatonin which is extracted into toluene and counted [126].

The Falck-Hillarp histofluoresence technique is not as sensitive for 5-HT as for the catecholamines but it has been used successfully for insect nervous tissue. Fluorescence can be increased by providing exogenous indoleamines which are concentrated by the cellular uptake mechanism [127].

4. Enzymes of Biosynthesis

Since the biogenic amines in insects serve as precursors for tanning reagents and pigments as well as having possible transmitter function, the distribution and time of maximum activity of the synthesizing enzymes is important in interpreting function. Murdock [128] has summarized evidence that there may be two pools of catecholamines, one outside the nervous system which fluctuates with the molt cycle and a neural pool which may have transmitter function. The activity of dihydroxyphenylalanine decarboxylase (dopa decarboxylase) which makes DA is particularly high in locust brain and other parts of the central nervous system. Activity elsewhere is more variable through the molt cycle. To demonstrate this enzyme a radiochemical method has been used [129] in which carboxyl-labeled dopa is decarboxylated to release labeled CO_2. Demonstration of the production of DA is also possible. This system, which shows little dependence on pyridoxal phosphate, did not decarboxylate L-tryptophan, L-histidine of L-phenylalanine. The presence of NA in the nervous system suggests that it may be a metabolite of DA. However, it could also be formed from tyramine via norsynephrine [128].

Methods for the study of vertebrate enzymes of catecholamine biosyn-thesis are found in reviews by Creveling and Daly [130] and Goldstein [131] and for hydroxy indole compounds in a review by Lovenberg and Engleman [122].

5. Inactivating Enzymes

In vertebrates the biogenic amines can be inactivated by MAO and COMT. Methods for detection of these enzymes have been reviewed by Jarrott [132].

In insects the presence of a mitochondrial MAO has been reported in some species but is apparently absent in others [128]. Although COMT is more important in inactivating catecholamines in vertebrates, it has not been studied in insects. However, Evans and Fox [133] report another inactivating mechanism: honeybee brain homogenates can N-acetylate indolealkylamines. They discovered this using an enzymatic method developed for vertebrate MAO in which the homogenate was incubated with (^{14}C) tryptamine. Products were separated and identified by thin layer chromatography and scintillation counting.

V. ABBREVIATIONS

ACh, acetylcholine; AChE, acetylcholinesterase; ChAc, choline acetylase =ChAT, choline acetyltransferase; ChE, cholinesterase; COMT, catechol-ortho-methyltransferase; DA, dopamine; GABA, γ-aminobutyric acid; GABA-T, GABA transaminase; GAD, glutamic acid decarboxylase; Glu, glutamic acid or gluta-mate; 5-HT, 5-hydroxytryptamine; MAO, monoamine oxidase; NE, norepinephrine; SSADH, succinic semialdehyde dehydrogenase.

VI. REFERENCES

1. J. J. Callec, in J. E. Treherne (Editor), Insect Neurobiology, North-Holland, Amsterdam, 1974, Ch. 3.

2. R. M. Pitman, Comp. Gen. Pharmacol., 2(1971) 347.

3. P. N. R. Usherwood and S. G. Cull-Candy, in P.N.R. Usherwood (Editor), Insect Muscle, Academic Press, London, 1975, Ch. 4

4. E. Florey, Federation Proc., 26 (1967) 1164.

5. H. M. Gerschenfeld, Physiol. Rev. 53 (1973) 1. .

6. D. Glick (Editor), Analysis of Biogenic Amines and Their Related Enzymes, Methods of Biochemical Analysis, Supplemental Vol., Interscience, New York, 1971.

7. L. L. Iversen, S. D. Iversen and S. H. Snyder (Editors), Biochemical Principles and Techniques in Neuropharmacology, Handbook of Psycho-pharmacology, Vol. 1, Plenum, New York, 1975.

8. N. Marks and R. Rodnight (Editors), Research Methods in Neuro-chemistry, Vol 1 and 2, Plenum, New York, 1972, 1974.

9. J. R. Cooper, F. E. Bloom and R. H. Roth, The Biochemical Basis of Neuropharmacology, 2nd Ed., Oxford, New York, 1974.

10. J. E. Smith, J. D. Lane, P. A. Shea, W. J. McBride and M. H. Aprison, Anal. Biochem. 64 (1975) 149.

11. D. J. Jenden and L. B. Campbell, in D. Glick (Editor), Analysis of Biogenic Amines and Their Related Enzymes, Interscience, New York, 1971, p. 183.

12. B. N. Smallman and C. A. Schuntner, Insect Biochem. 2 (1972) 67.

13. J. E. Treherne and D. S. Smith, J. Exp. Biol. 43 (1965) 13.

14. F. Fonnum, Biochem. Pharmacol. 17 (1968) 2503.

15. L. T. Potter and W. Murphy, Biochem. Pharmacol. 16 (1967) 1386.

16. R. Massarelli, A. Ebel and P. Mandel, Anal. Biochem. 57 (1974) 299.

17. G. Zweig and J. R. Whitaker, Paper Chromatography and Electrophoresis, Vol. 1, Academic Press, New York, 1967, p. 326.

18. Y. Dunant and L. Hirt, J. Neurochem. 26 (1967) 657.

19. F. C. MacIntosh and W. L. M. Perry, Methods in Medical Research, 3 (1950) 78.

20. A. Ahmed and N. R. W. Taylor, J. Pharm. Pharmacol., 9 (1957) 536.

21. J. H. Welsh and B. Twarog, Methods in Medical Research, 8 (1960) 187.

22. T. Forrester, J. Physiol., 187 (1966) 12 p.

23. J. H. Fellman, J. Neurochem., 16 (1969) 135.

24. D. J. Jenden, M. Roch and R. A. Booth, Anal. Biochem., 55 (1973) 438.

25. S. J. Fidone, S. T. Weintraub and W. B. Stavinoha, J. Neurochem., 26 (1976) 1047.

26. E. H. Colhoun, Nature, 182 (1958) 1378.

27. R. E. McCaman and J. M. Hunt, J. Neurochem. 12 (1965) 253.

28. F. Fonnum, Biochem. J., 115 (1969) 465.

29. P. C. Emson, M. Burrows and F. Fonnum, J. Neurobiol., 5 (1974) 33.

30. A. B. Burt, J. Histochem. Cytochem., 18 (1970) 408.

31. P. Kasa, S. P. Mann and C. Hebb, Nature 226 (1970) 814.

32. J. Rossier, Brain Res., 98 (1975) 619.

33. G. I. Franklin, J. Neurochem., 26 (1976) 639.

34. A. Silver, The Biology of Cholinesterases, Elsevier, Amsterdam, 1974.

35. K. B. Augustinsson, in D. Glick (Editor), Analysis of Biogenic Amines and Their Related Enzymes, Interscience, New York, 1971. p. 127.

36. B. W. Smallman and A. Mansingh, Annual Review of Entomology 14 (1969) 387.

37. D. A. Wustner and R. T. Fukuto, J. Agr. Food Chem., 21 (1973) 756.

38. H. H. Moorefield, Contrib. Boyce Thompson Inst., 18 (1957) 463.

39. M. Habibulla and R. W. Newburgh, Insect Biochem., 3 (1973) 231.

40. C. T. Huang and W. C. Dauterman, Insect Biochem., 3 (1973) 325.

41. R. K. Tripathi and R. D. O'Brien, Pesticide Biochem. Physiol., 3 (1973) 495.

42. A. P. Beranek, Ent. Exp. Appl. 17 (1974) 129.

43. K. B. Augustinsson, in D. Glick (Editor), Methods of Biochemical Analysis, Vol. 5, Interscience, New York, 1957, p. 1.

44. L. E. Chadwick, J. B. Lovell and V. E. Egner, Biol. Bull., 104 (1953) 323.

45. J. L. Brik and Y. E. Mandel'shtam, J. Evol. Biochem. Physiol., 9 (1973) 120.

46. G. L. Ellman, K. D. Courtney, V. Andres, Jr. and R. M. Featherstone, Biochem. Pharmacol., 7 (1961) 88.

47. S. Hestrin, J. Biol. Che., 180 (1949) 249.

48. F. P. W. Winteringham and R. W. Disney, Biochem. J., 91 (1964) 506.

49. S. Koslow and E. Giacobini, J. Neurochem., 16 (1969) 1523.

50. J. G. Hildebrand, J. G. Townsel and E. A. Kravitz, J. Neurochem., 23 (1974) 951.

51. G. B. Koelle, J. Pharm. Exp. Therap., 100 (1950) 158.

52. A.-H. Lee, R. L. Metcalf and G. M. Booth, Ann. Entomol. Soc. Am., 66 (1973) 333.

53. U. E. Brady, Ent. Exp. Appl., 13 (1970) 423.

54. A. Hess, Brain Res., 46 (1972) 287.

55. N. Frontali, R. Piazza and R. Scopelliti, J. Insect Physiol., 17 (1971) 1833.

56. A. M. Burt and A. Silver, Brain Res. 57 (1973) 518.

57. G. M. Booth and G. S. Whitt, Tissue & Cell, 2 (1970) 521.

58. R. E. Papka, Cell Tissue Res. 162 (1975) 185.

59. H. C. Bauer, J. Insect Physiol., 22 (1976) 683.

60. V. B. Wigglesworth, Quart. J. Micros. Sci., 99 (1958) 441.

61. M. J. Karnovsky, J. cell Biol., 23 (1964) 217.

62. R. J. Barrnett, J. Cell Biol., 12 (1962) 247.

63. D. S. Smith and J. E. Treherne, J. Cell Biol., 26 (1965) 445.

64. M. K. Gaitonde, in N. Marks and R. Rodnight (Editors), Research Methods in Neurochemistry, Vol. 2, Plenum Press, New York, 1974, p. 321.

65. G. M. Holman and B. J. Cook, J. Insect Physiol., 16 (1970) 1891.

66. I. R. Faeder and M. M. Salpeter, J. Cell Biol., 46 (1970) 300.

67. P. N. R. Usherwood, in J. E. Treherne (Editor), Insect Neurobiology, North-Holland, Amsterdam, 1974, p. 245.

68. N. Frontali and R. Pierantoni, Comp. Biochem. Physiol., 44A (1973) 1369.

69. J. A. Campos-Ortega, Z. Zellforsch. Mikroskop. Anat. 147 (1974) 415.

70. I. A. Sytinsky, B. M. Guzikov, M. V. Gomanko, V. P. Eremin and N. N. Knovalova, J. Neurochem., 25 (1975) 43.

71. G. A. Kerkut, L. D. Leake, A. Shapira, S. Cowan and R. J. Walker, Comp. Biochem. Physiol., 15 (1965) 485.

72. V. J. Balcar and G. A. R. Johnson, J. Neurochem., 19 (1972) 2657.

73. M. H. Aprison, W. J. McBride and A. R. Freeman, J. Neurochem., 21 (1973) 87.

74. W. J. McBride, A. R. Freeman, L. T. Graham, Jr., and M. H. Aprison, Brain Res., 59 (1973) 440.

75. P. N. R. Usherwood, P. Machili and G. Leaf, Nature, 219 (1968) 1169.

76. A. Daoud and R. Miller, J. Neurochem., 26 (1976) 119.

77. R. A. Yates and P. Keen, Brain Res. 107 (1976) 117.

78. L. T. Graham and M. H. Aprison, Anal. Biochem., 15 (1966) 487.

79. E. A. Kravitz and D. D. Potter, J. Neurochem., 12 (1965) 323.

80. S. J. Enna and S. H. Snyder, J. Neurochem., 26 (1976) 221.

81. J. F. Donnellan, D. W. Jenner and A. Ramsay, Insect Biochem., 4 (1974) 243.

82. G. W. O. Oliver, P. V. Taberner, J. T. Rick and G. A. Kerkut, Comp. Biochem. Physiol., 38B (1971) 529.

83. P. Langcake and A. N. Clements, Insect Biochem., 4 (1974) 225.

84. N. Frontali, Nature, 191 (1961) 178.

85. P. M. Fox and J. R. Larsen, J. Insect Physiol. 18 (1972) 439.

86. H. A. Robertson, J. Exp. Biol. 63 (1975) 413.

87. E. Östlund, Acta Physiol. Scand. Suppl. 112, 1954, p. 1.

88. U. S. von Euler, Nature, 190 (1961) 170.

89. N. Frontali and J. Häggendal, Brain Res., 14 (1969) 540.

90. J. K. Koeppe and R. R. Mills, Insect Biochem., 5 (1975) 399.

91. N. J. Lane, in J. E. Treherne (Editor), Insect Neurobiology, North-Holland, Amsterdam, 1974, p. 1.

92. G. Mancini and N. Frontali, Z. Zellforsch. Mikroskop. Anat., 103 (1970) 341.

93. M. Gersch, E. Hentschel and J. Ude, Zool. Jahrb. Abt. Allgem. Zool. Physiol. Tiere, 78 (1974) 1.

94. T. A. Miller, in P. N. R. Usherwood (Editor), Insect Muscle, Academic Press, London, 1975, p. 545.

95. N. Klemm and L. Schneider, Comp. Biochem. Physiol., 50 (1975) 177.

96. W. Feldberg and C. H. Kellaway, Australian J. Exp. Biol. Med. Sci., 15 (1937) 461.

97. J. Glowinski, I. J. Kepin and J. Axelrod, J. Neurochem., 12 (1965) 25.

98. H. Weil-Malherbe, in D. Glick (Editor), Analysis of Biogenic Amines and Their Related Enzymes, Interscience, New York, 1971, p. 119.

99. S. H. Snyder and K. M. Taylor, in N. Marks and R. Rodnight (Editors), Research Methods in Neurochemistry, Vol. 1, Plenum Press, New York, 1972, p. 287.

100. N. N. Osborne, Microchemical Analysis of Nervous Tissue. Pergamon Press, Oxford, 1974.

101. K. D. McMurtrey, L. R. Meyerson, J. L. Cashaw and V. E. Davis, Anal.
     Biochem. 72 (1976) 566.

102. F. P. Abramson, M. W. McCaman and R. E. McCaman, Anal. Biochem.,
     57 (1974) 482.

103. U. S. von Euler, Pharmacol. Rev., 11 (1959) 262.

104. J. Häggendal, Pharmacol. Rev., 18 (1966) 325.

105. H. C. Campuzano, J. E. Wilkerson and S. M. Horvath, Anal. Biochem.,
     64 (1975) 578.

106. R. Laverty and K. M. Taylor, Anal. Biochem., 22 (1968) 269.

107. J. G. Hildebrand   D. L. Barker, E. Herbert and E. A. Kravitz, J.
     Neurobiol., 2 (1971) 231.

108. M. A. Beaven, in L. L. Iversen, S. D. Iversen and S. H. Snyder
     (Editors),Handbook of Psychopharmacology, Vol. 1, Plenum Press, New
     York, 1975, p. 253.

109. A. C. Cuelle, R. Riley and L. L. Iversen, J. Neurochem., 21 (1973)
     1337.

110. J. T. Coyle and D. T. Henry, J. Neurochem. 21 (1973) 61.

111. P. G. Passon and J. D. Peuler, Anal. Biochem., 51 (1973) 618.

112. C. Gauchy, J. P. Tassin, J. Glowinski and A. Cheramy, J. Neurochem.,
     26 (1976) 471.

113. B. Falck and C. Owman, Acta Univ. Iund, 1965, Sect. II, No. 7, p. 1

114. A. Björklund, B. Falck and C. Owman, in J. E. Rall and I. Kopin
     (Editors), Methods of Investigative and Diagnostic Endocrinology,
     North-Holland, Amsterdam, 1972, p. 318.

115. A. Björklund, B. Falck and N. Kelmm, J. Insect Physiol., 16 (1970)
     1147.

116. N. Frontali, J. Insect Physiol., 14 (1968) 881.

117. O. Lindvall and A. Björklund, Histochem., 39 (1974) 97.

118. A. A. Boulton and J. R. Majer, in N. Marks and R. Rodnight (Editors)
     Research Methods in Neurochemistry, Vol. 1, 1972, p. 341.

119. P. Molinoff and J. Axelrod, Science, 164 (1969) 428.

120. J. M. Saavedra, Anal. Biochem., 59 (1974)628.

121. A. J. Harmar and A. S. Horn, J. Neurochem., 26 (1976) 987.

122. S. W. Lovenberg and K. Engelman, In D. Glick (Editor), Analysis of
     Biogenic Amines and Their Related Enzymes, Interscience, New York,
     1971, p. 1.

123. J. H. Welsh and B. Twarog, in H. D. Bruner (Editor), Methods in
     Medical Research, Vol. 9, Year Book Publishers, Chicago, 1960, p. 187.

124. D. F. Bogdanski, A. Pletscher, B. B. Brodie and S. Udenfriend, J. Pharm. Exp. Therap., 117 (1956) 82.

125. S. H. Snyder, J. Axelrod, and M. Zweig, Biochem. Pharmacol., 14 (1965) 831.

126. J. M. Saavedra, M. Brownstein and J. Axelrod, J. Pharmacol. Exp. Therap. 186 (1973) 508.

127. N. Klemm and L. Schneider, Comp. Biochem. Physiol., 50C (1975) 177.

128. L. L. Murdock, Comp. Gen. Pharmacol., 2 (1971) 254.

129. L. L. Murdock, R. A. Wirtz and G. Köhler, Biochem. J., 132 (1973) 689.

130. C. R. Creveling and J. W. Daly, in D. Glick (Editor), Analysis of Biogenic Amines and Their Related Enzymes, Interscience, New York, 1971, p. 153.

131. M. Goldstein, in N. Marks and R. Rodnight(Editors), Research Methods in Neurochemistry, Vol. 1, Plenum, New York, 1952, p. 317.

132. B. Jarrett, in N. Marks and R. Rodnight (Editors), Research Methods in Neurochemistry, Vol. 2, Plenum, New York, 1974, p. 377.

133. P. H. Evans and M. P. Fox, J. Insect Physiol., 21 (1975) 343.

SUBJECT INDEX